Lecture Notes in Statistics

Edited by J. Berger, S. Fienberg, J. Gani,
K. Krickeberg, I. Olkin, and B. Singer

60

Lázló Györfi Wolfgang Härdle
Pascal Sarda Philippe Vieu

Nonparametric Curve Estimation from Time Series

Springer-Verlag
Berlin Heidelberg New York London Paris Tokyo Hong Kong

Authors

Lázló Györfi
Hungarian Academy of Sciences, Technical University Budapest
Stoczek u. 2., 1521 Budapest, Hungary

Wolfgang Härdle
Rechts- und Staatswissenschaftliche Fakultät
Wirtschaftstheoretische Abteilung II, Universität Bonn
Adenauerallee 24–26, 5300 Bonn, Federal Republic of Germany

Pascal Sarda
Philippe Vieu
Laboratoire de Statistique et Probabilités
Unité associée C.N.R.S. 745, Université Paul Sabatier
118, Route de Narbonne, 31062 Toulouse Cédex, France

Mathematical Subject Classification: 62G05, 62M10

ISBN 0-387-97174-2 Springer-Verlag New York Berlin Heidelberg
ISBN 3-540-97174-2 Springer-Verlag Berlin Heidelberg New York

Printed in Germany

Printing and binding: Druckhaus Beltz, Hemsbach/Bergstr.
2847/3140-543210 – Printed on acid-free paper

dédié à

Gérard Collomb,
Marie Paule,
Etienne et Thibault.

PREFACE.

This book constructs bridges between two major themes in mathematical statistics: Time Series and Smoothing. Both themes have a long history that reaches back at least to the 19th century. In the last century statistics has been used mostly in an empirical way. In the last seventy years, however, emphasis of statistics has been more on the theoretical properties of time series and curve estimation. As a consequence both themes have been intensively studied from a mathematical angle of view.

Starting in the twenties, time series have been considered mainly from the parametric viewpoint. This viewpoint has the advantage that if the observed time series is sufficiently described by a parametric model, then a relatively small set of (structural) parameters serves as tools for interpretation and inference. A disadvantage of this viewpoint is a statistical situation where the observed data do not follow a specific parametric model. A consequence of imposing such a non-appropriate parametric model results in a bias that dominates the statistical error asymptotically.

This misspecification problem can be overcome by nonparametric smoothing techniques. It is the strength of the nonparametric approach to consider a richer collection of models and functions with general shape rather than a relatively small set of parametrised curves.

The mathematical analysis of smoothing operations reaches back to the fifties. In the setting of curve estimation from a sequence of independent identically distributed variables, Rosenblatt (1956) and Parzen (1962) have used kernel estimation methods for approximating a density function. In their setting they assumed the probability structure of the observations to be i.i.d.. Later Bartlett (1963) extended these smoothing ideas for estimating a spectral density from a second order stationary time series. The last decade has seen an enormeous increase of interest in nonparametric curve estimation. Huber (1983) mentions that some statisticians attribute this to the increased computing power. The nowadays available "low cost computing power" makes parameterfree smoothing algorithms efficient and interactive. We believe that not the computer alone is the driving force behind the interest in smoothing techniques but rather the insight of many statisticians that the misspecification of a model should be avoided. On the other hand we have to pay a price for using the parameterfree approach: the curve, since it stems from a richer collection of models, is less accurate (harder) to estimate. Nevertheless we find it important to study the approximation of curves over a wide class of functions even if we do not achieve the same asymptotic efficiency as for a narrower class of parametric models.

In time series analysis smoothing problems occur of course in the spectral domain when we want to estimate the spectral density e.g. for model fitting. In the time domain nonparametric prediction is one of the fields where smoothing methods are intensively used. An example is the waterflow prediction from a time series of river data. It is also interesting to estimate the marginal density of a stationary time series. The marginal density is the predictive density for long time forecasting. Nonparametric forecasting of economic time series such as stocks or gold prices have been recently studied. Estimation of hazard functions is a standard tool in analysis of clinical survival data. It is the purpose of this book to study these nonparametric smoothing techniques for time series and to provide mathematical tools for nonparametric estimation under general dependence assumptions.

Gérard Collomb has consequently applied the parameterfree thinking to time series prediction. His work has been very stimulative for all of us. In the last three years we have had particularly collaboration with Adrian Bowman, Ray Carroll, Henri Caussinus, Luc Devroye, Jeff Hart, Kurt and Werner Hildenbrand, Steve Marron, David Scott and Sid Yakowitz. Their ideas and criticisms shaped this book in an essential way. We would like to thank them for their helpful cooperation.

The second author has given several short courses on the subject in Université des Sciences Sociales, Toulouse; G.R.E.Q.E., Marseille and Universidad de Santiago, Santiago de Compostella. It is a pleasure to thank collegues and students at these places.

Finally, we gratefully acknowledge the financial support of the Deutsche Forschungsgemeinschaft, the French Ministery of Education, the French Centre National pour la Recherche Scientifique, and the Hungarian National Foundation for Scientific Research.

Lázló Györfi
Wolfgang Härdle
Pascal Sarda
Philippe Vieu.

Table of contents.

Chapter VI: HOW TO SELECT THE SMOOTHING PARAMETER

Chapter I

INTRODUCTION.

A question that is quite often asked in time series analysis is the one after the value of future observations. This is a very frequent question in economics. One wants to know whether some regularly observed economic indicator follows certain regularities. Once a regularity is detected, prediction and also speculation can be based on it. This is probably the most striking example that most people are aware of.

Of course there are more prediction problems than just this economic one. In medicine, dike and water ressources management, prediction of future values of time series are important tools for planning and inference. In medical monitoring, for example, certain threshold values indicate immediate actions. When is a time series of monitored medical data about to break through such a threshold? This can be answered through inference on the predicted values. In water ressources management, one would like to predict waterflow from certain rivers in order to generate steady water supply for certain regions. It is quite important for many people to have reservoirs filled up when a dry period is expected. Also it is vital to know whether dikes may be flooded.

A common statistical approach for solving such problems is to postulate a certain parametric model. In the present context one possible parametric model could be a linear autoregressive process. Thus the prediction is based on the finite set of estimated parameters. The parameterfree, nonparametric approach to this problem allows the prediction curve to be any smooth function. Thus the prediction is based on the knowledge of a whole function. The basic difference between these two approaches lies in the fact that for the latter we have to compute a whole function instead of a set of parameters.

In order to highlight the differences between these two approaches let us consider the prediction problem in a little more detail. Assume that a time series $(Z_i)_N$ is given of which we have observed n data points. Suppose we wish to predict Z_{n+1} from Z_n.

Parametric thinking. The statistician leaning towards parametric modelling could have reasons to build his model such that

$$Z_{i+1} = \alpha Z_i + \epsilon_i,$$

i.e. the process is linear autoregressive with parameter α and innovation errors (ϵ_i). If we consider a scatterplot of the pairs (Z_i, Z_{i+1}), such an approach is justified if a straight line through the origin is clearly visible. The prediction thus would be

$$\hat{Z}_{n+1} = \hat{\alpha} Z_n.$$

<u>Parameterfree thinking.</u> The statistician who has reasons to believe that the straight line in the scatterplot of the pairs (Z_i, Z_{i+1}) insufficiently describes the relation between Z_i and Z_{i+1}, would probably try a smoothing approach. Typically such a function is estimated by a local average (in space, <u>not</u> in time), i.e.

$$\hat{r}(Z_n) = n^{-1} \sum_{i=1}^{n} W_{n,i}(Z_i; Z_1, \ldots, Z_n) Z_i$$

serves as a predictor for Z_{n+1}. Here $W_{n,i}(.)$ denotes a weight sequence depending on the past $(Z_i : i \le n)$. Special forms of weights $W_{n,i}$ are given by kernel functions to be described below in chapter 3.

Both approachs are compared in the following example from Yakowitz (1985). Dealing with the flood warning problem of rivers, the author applies an ARMA-type predictor and a nonparametric one to predict flow Z_{n+1}. Figure 1.1 shows that both methods produce the same overall prediction curve. Nevertheless, the nonparametric approach produces a predictor with smaller bias for the peak flow levels.

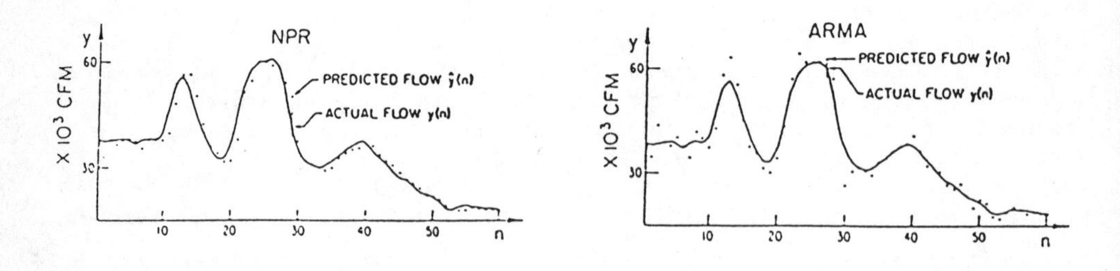

Figure 1.1. Kotenaii river flow prediction. Nonparametric and parametric fits against the actual flows (n=60). ARMA=parametric fit by an ARIMA (1,1,2) model; NPR=nonparametric fit by the kernel predictor.

In order to make this averaging process work well in an asymptotic sense, certain assumptions on the dependence structure of the time series are to be imposed. We will work mainly with three mixing conditions:

φ-mixing,

ρ-mixing,

α-mixing.

We will also consider some variants of these three mixing assumptions, and investigate the relationships in between them. This is done in chapter 2. The mathematical reader familiar with these conditions might want to skip this section.

In chapter 3, the problem of predicting future values Z_{n+s} from past values is investigated, via the estimation of the autoregression functions, as a particular case of the more general problem of prediction for a bivariate time series (X_i, Y_i). Nonparametric kernel smoothers of the regression function of Y on X

$$r(.) = E(Y|X=.),$$

are defined. First, universal consistency properties are investigated and uniform convergence results are stated. These results apply directly to the problem defined above, i.e. the prediction of future values of time series by the kernel smoother $\hat{r}$.

Another tool to get insight into the structure of a stationary time series (Z_i) is to estimate the common density function f of Z_i. If the variables are not i.i.d. then the marginal density does not characterize the probability law of Z_i. If we want to estimate the density of $X=(Z_1,\ldots,Z_d)$ then for given Z_i we can have the "sliding window" sample $X_1, X_2, \ldots$ where $X_i=(Z_i, Z_{i+1}, \ldots, Z_{i+d-1})$ which is obviously dependent even for i.i.d. (Z_i). In this setting also, it is quite interesting to investigate parameterfree methods, since in practical situations the function f does not necessarly belong to some parametric class. Nonparametric density estimates have been dominantly investigated in the setting of independent data (see the surveys by Devroye and Györfi (1984) and Silverman (1986)). The aim of chapter 4 is to investigate the behavior of some nonparametric estimates of f when the time series (Z_i) satisfies some mixing conditions. In this chapter we adopt the L_1 view of Devroye and Györfi (1984)

The problem of distribution estimation is then adressed in chapter 5. Generalisations of Glivenko-Cantelli's theorem are derived under mixing conditions using analytic arguments. The hazard function defined as

$$\lambda(.) = \frac{f(.)}{1-F(.)},$$

where f and F stand for the density and the distribution function of a stationary time series $(Z_i)_N$ is estimated by a ratio of an estimate of f over an estimate of 1-F. The empirical distribution function is used to estimate F while the density is estimated using kernel or k-NN estimate. Uniform convergence results are derived for those hazard estimates, showing that their asymptotic behavior are mainly influenced by the density estimate.

Finally, in chapter 6 we investigate an important practical problem. All the methods we consider depend on a smoothing parameter that determines the rate of the estimator. Theoretical studies in chapter 3,4,5 give a rate for this smoothing parameter but leave the question open of how to select it in practice. We investigate several automatic methods for finding the smoothing parameter. Data-driven selectors are defined for each of the three estimation problems, i.e. regression, density and hazard estimation, and asymptotic optimality results are shown when the variables are α-mixing. Examples are given that show the applicability of the proposed method.

Chapter II

2. DEPENDENT SAMPLES.

2.1. Stationary and ergodic sequences.

Let $(Z_i)_N$ be a sequence of random variables taking values in a metric vector space (either $\mathbb{R}^d$ or a Banach space B) and F_n^m (for n and m $\in$ Z U $\{-\infty,+\infty\}$) be the σ-algebra generated by $\{Z_i, n \le i \le m\}$. In this chapter we summarize some dependence structures and some basic results which are used in the sequel. If we say that $(Z_i)_N$ is *stationary*, then it means strict stationarity. For the definition of ergodicity we refer to Ibragimov and Linnik (1971).

Theorem 2.1.1. (Beck, 1963). If $(Z_i)_N$ takes values in a separable Banach space, is stationary and ergodic, and

$$E \| Z_1 \| < 1$$

then

$$\lim_n \| \frac{1}{n} \sum_{i=1}^{n} Z_i - EZ_1 \| = 0 \text{ a.s.}$$

Beck (1963) gave a proof of this theorem for i.i.d. $(Z_i)_N$, but the proof can be extended to the framework of stationary and ergodic $(Z_i)_N$.

We start with some definitions of mixing random variables and give basic tools for the different mixing types.

2.2. Mixing sequences.

2.2.1. φ-mixing sequences.

Definition 2.2.1. (Ibragimov, 1962). We say that $(Z_i)_N$ is φ-mixing (or uniformly mixing) if for the mixing coefficients

$$\varphi_k = \sup_n \sup_{\substack{A \in F_{-\infty}^n \\ P(A)>0 \\ B \in F_{n+k}^{\infty}}} |P(B \mid A)-P(B)|,$$

we have

$$\lim_{k\to\infty} \varphi_k=0.$$

A fundamental tool used to show asymptotic results is the following inequality due to Gerard Collomb. This exponential inequality is an extension to φ-mixing variables of the well-known Bernstein inequality (Hoeffding, 1963). It will be used very frequently in the sequel for results concerning φ-mixing variables.

Theorem 2.2.1. Collomb's inequality. (Collomb, 1984). If $(Z_i)_N$ is φ-mixing with

$$EZ_i = 0, \ |Z_i| \leq d, \ EZ_i^2 \leq D \text{ and } E|Z_i| \leq \delta,$$

then we have

$$P(|\sum_{i=1}^{n} Z_i| > \epsilon) \leq \exp\,(3\sqrt{e}n\frac{\varphi_m}{m} - \alpha\epsilon + \alpha^2 n6(D+4\delta d \sum_{i=1}^{m} \varphi_i)),$$

where α is a real and m an integer satisfying

$$1 \leq m \leq n \text{ and } \alpha md \leq 1/4.$$

It is worth mentioning the papers by Bradley (1980a) and Collomb and Doukhan (1983) who provide additional probability inequalities about φ-mixing sequences. These results are very interesting but leave somewhat the scope of this book.

2.2.2. ρ-mixing sequences.

Definition 2.2.2. (Kolmogorov and Rozanov, 1960). We say that $(Z_i)_N$ is ρ-mixing (or asymptotically uncorrelated) if for the mixing coefficients

$$\rho_k = \sup_n \sup_{\substack{X \in L_2(F_{-\infty}^n) \\ Y \in L_2(F_{n+k}^{\infty})}} \frac{E[(X-EX)(Y-EY)]}{[Var(X)Var(Y)]^{1/2}},$$

we have

$$\lim_{k\to\infty} \rho_k = 0.$$

Here $L_2(F_n^m)$ stands for the set of random variables X which are measurable on F_n^m and such that $0 < EX^2 < \infty$.

By slight modifications of the techniques used by Collomb to prove theorem 2.2.1 above, M. Peligrad obtained the following exponential inequality for ρ-mixing variables.

Theorem 2.2.2. (Peligrad, 1988, Corollary 3.4). Let $(Z_i)_N$ be a sequence of ρ-mixing variables such that

$$EZ_i = 0, \ |Z_i| \leq d \ \text{ and } E(\sum_{i=1}^{n} Z_i)^2 \leq D.$$

We have

$$P(\sum_{i=1}^{n} Z_i > \epsilon) \leq U(\mu) \exp(-c_1\epsilon/[8nd+D(c_2-(2nr^{-1/2})\rho_n(r))^{-1/2}]),$$

where μ is a real number and r an integer such that

$$\mu > 1 \text{ and } r \leq n/2,$$

and where

$$U(\mu)=[(2\mu-1)^{-1/2}-(2\mu-1)^{-1}]^{-1}, \ c_1=\log(2\mu-1), \ c_2=4(10\mu-1)^{-1}$$

and

$$\rho_n(r) = \max_{1\leq k\leq n-r} \rho(F_0^k, F_{k+r}^n),$$

F_i^j being, for $i<j$, the σ-algebra spanned by $(Z_i,\ldots,Z_j)$ and

$$\rho(F,G) = \max_{X\in F, Y\in G} |E(XY) - E(X)E(Y)|/(\text{Var}X \ \text{Var}Y)^{1/2}.$$

Theorem 2.2.3. (Peligrad, 1988, Proposition 3.4). Let $(Z_i)_N$ be a sequence of ρ-mixing variables such that

$$EZ_i = 0 \text{ and } EZ_i^2 < \delta,$$

then we have

$$E\left(\sum_{i=1}^{n} Z_i\right)^2 \le n d_n \delta,$$

where

$$d_n = 8000\exp\{3 \sum_{i=1}^{[\log n]} \rho_n^*(2^i)\},$$

and

$$\rho_n^*(j) = \max_{1\le k\le n-j-1} \max_{1\le m\le (n-j-k)/2} \left| E\left(\sum_{i=k}^{k+m} Z_i, \sum_{i=k+m+j}^{k+2m+j} Z_i \right) \right|.$$

Let us also mention that Bradley (1980b and 1981a) gives related results about ρ-mixing sequences.

2.2.3. α-mixing sequences.

Definition 2.2.3. (Rosenblatt, 1956). We say that $(Z_i)_N$ is α-mixing (or strong mixing) if for the mixing coefficients

$$\alpha_k = \sup_n \sup_{\substack{A\in F_{-\infty}^{n} \\ B\in F_{n+k}^{\infty}}} |P(A \cap B)-P(A)P(B)|,$$

we have

$$\lim_{k\to\infty} \alpha_k = 0.$$

Theorem 2.2.4. (Yokoyama, 1980). If $(Z_i)_N$ is an α-mixing sequence of 0 mean and bounded random variables and if for some $\delta_1>0$

$$\sum_{i=1}^{U} i^{\delta_1}\alpha_i < \infty$$

then there is a constant K such that

$$E\left|\sum_{i=1}^{n} Z_i\right|^{2+2\delta_1} \le K n^{1+\delta_1}.$$

Theorem 2.2.5. (Dehling and Philipp, 1982). If $(Z_i)_N$ is α-mixing, X and Y are random variables taking values in a separable Hilbert space, and if X is measurable on $F_{-\infty}^{n-m}$ and Y is measurable on F_n^{∞}, moreover if

$$\|X\| \le 1, \ \|Y\| \le 1 \text{ a.s.},$$

then

$$|E(X,Y)-(EX,EY)| \le 10\alpha_m.$$

This result was also proven by Hall and Heyde (1980, theorem A5). Note also that theorem 6.2.1 below is an extension, to a very special case, of this theorem 2.2.5.

Following the line of proof of Collomb (1984), Carbon (1983) obtained the following exponential inequality for α-mixing variables. (See also the Carbon, 1982.)

Theorem 2.2.6. (Carbon, 1983). If $(Z_i)_N$ is α-mixing with

$$EZ_i = 0, \quad |Z_i| \le d, \text{ and } EZ_i^2 \le D,$$

then we have

$$P(|\sum_{i=1}^{n} Z_i|>\epsilon) \le 2\exp(-\alpha\epsilon+6\alpha^2 e(D+8d^2 \sum_{i=1}^{s} \alpha_i)n+2\sqrt{e}\alpha_s^{2s/5n} ns^{-1})$$

where α is a real and s an integer satisfying

$$1 \le s \le n \text{ and } 0 \le \alpha \le (sde)/4.$$

For further discussion on α-mixing variables see Bradley (1980b and 1981a).

2.2.4. Examples and general comments.

How are the mixing conditions related to each other ?

It is known that

$$4\alpha_k \le \rho_k \le 2(\varphi_k)^{1/2} \text{ and } \alpha_k \le \varphi_k \text{ for all k.}$$

Ibragimov and Linnik (1971) have shown, that for Gaussian process $\{Z_i\}$ the φ-mixing condition is equivalent to m-dependence, which means that φ-mixing implies that there is an integer m>0 such that

$$\varphi_k = 0 \text{ for all } k \ge m.$$

Moreover for Gaussian processes we have (Kolmogorov and Rozanov, 1960)

$$\alpha_k \le \rho_k \le \alpha_k 2\pi,$$

thus for Gaussian process α-mixing and ρ-mixing are equivalent.

The majority of papers, dealing with density estimation from dependent samples, considers the case of Markov process. An often used assumption there is **Doeblin's condition**. Doeblin's condition is stated in exact form in (3.4.1) below. If a stationary Markov process satisfies Doeblin's condition then it is φ-mixing and there are constants $0 < A < \infty$ and $0 < a < 1$ such that

$$\varphi_k \leq Aa^k,$$

thus it is exponentially φ-mixing (Rosenblatt, 1971, p.209).
The reader will find details about this condition in Doob (1953) and in section III.4.3 below. It is interesting to note that if a stationary Markov process is φ-mixing then it is exponentially φ-mixing (Davydov, 1973). Note that an alternative condition for a stationary Markov process is the so called G_2 condition, which implies that the process is exponentially α-mixing (Rosenblatt, 1971b, p.200), see also Yakowitz (1988).

Linear Processes

Mixing properties are most easily studied for linear processes,

$$Z_n = \sum_{I=0}^{\infty} g_i Y_{n-i} \text{ and } g_0 \neq 0,$$

where Y_i are i.i.d. The mixing properties of Z_n can be derived from the summability conditions of (g_i) (Withers, 1981). Chanda (1974) gave conditions on (g_i) which insure that such a process is α-mixing. Gorodetskii (1977) improved and partially corrected these results.

A non stationary φ-mixing process

A special case of a process not necessarily stationary that satisfies a mixing condition is the autoregressive process

$$Z_i = R(Z_{i-1},\dots,Z_{i-q}) + \epsilon_i,$$

q being a nonnegative integer and (ϵ_i) a sequence of i.i.d variables with zero mean. Doukhan and Ghindes (1980a) presented conditions under which such a process is φ-mixing. These conditions may be found in section III.4.4 below which is particularly devoted to the study of such processes.

The mixing conditions we study

Along this book we shall only deal with the three kinds of mixing conditions defined above, i.e.,

φ-mixing,
ρ-mixing,
α-mixing,

since they are the most often used in practice and alos because most of the other existing mixing conditions are very close to one of these three. We refer to the survey by Bradley (1985) for a more exhaustive list of mixing conditions. The survey by Doukhan and Leon (1988) and chapter 5 in Hall and Heyde are also very useful for that. We mention that further mixing conditions are defined and studied by Bradley (1980b,c and d).

2.3. Martingale differences and mixingale.

Definition 2.3.1. The process $(Z_i)_N$ is called a martingale difference if it is real valued and with $N^* = N \cup \{0\}$

$$E(Z_n | F_{-\infty}^{n-1}) = 0 \text{ for } n \in N^*.$$

Theorem 2.3.1. (Azuma, 1967). If $(Z_i)_N$ is a martingale difference and

$$|Z_i| \leq K \text{ a.s.},$$

then for all $\epsilon > 0$

$$P(|\sum_{i=1}^{n} Z_i| > \epsilon) \leq 2e^{-\frac{\epsilon^2}{2nK^2}}.$$

Theorem 2.3.2. If $(Z_i)_N$ is real valued and

$$|Z_i| \leq K \text{ a.s.},$$

then for all integer $m > 0$ and all $\epsilon > 0$

$$P(|\sum_{i=1}^{n} (Z_i - E(Z_i | F_{i-m}))| > \epsilon) < 2m\, e^{-\frac{\epsilon^2}{2nm^2K^2}}.$$

Proof of theorem 2.3.2. We have

$$P(|\sum_{i=1}^{n} (Z_i - E(Z_i | F_{i-m}))| > \epsilon)$$

$$\leq \sum_{j=1}^{m} P(|\sum_{i=1}^{n} (E(Z_i | F_{i-j+1}) - E(Z_i | F_{i-j}))| > \epsilon/m).$$

For each fixed j, the process $\{E(Z_i | F_{i-j+1}) - E(Z_i | F_{i-j}),\ -\infty < i < \infty\}$ forms a martingale difference. Therefore the theorem follows from theorem 2.3.1.

McLeish (1974) introduced the mixingale as the generalization of a martingale difference, and extended the stability theorems of martingale differences. Here we use a simplified version of a mixingale. We also consider Hilbert space valued mixingales. Let H denote a Hilbert space with inner product $(.,.)$ and norm $\|.\|$.

Definition 2.3.2. We say that $\{(Z_n, F_n),\ -\infty < n < \infty\}$ is an adapted mixingale if Z_n is H valued random variable which is measurable with respect to the

σ-algebra F_n $(n\geq 1)$ and

a. $E(Z_n)=0$ for $n\geq 1$,

b. $F_n=\{\phi, \Omega\}$ for $n\leq 0$ and $F_n \subset F_{n+1}$ for $n\geq 1$,

c. There are sequences of constants $(c_n, n\in N^*)$ and $(\psi_m, m\in N)$ such that

$$E\|E(Z_n|F_{n-m})\|^2 \leq \psi_m^2 c_n^2 \text{ for all } n\geq 1 \text{ and } m\geq 0.$$

__Theorem 2.3.3.__ (Györfi and Masry, 1988). Let $\{(Z_n, F_n)\}$ be a simple mixingale such that

$$\sum_{j=1}^{\infty} c_j^2 < \infty \tag{2.3.1}$$

and such that for some $\delta>0$

$$\sum_{n=1}^{\infty} (\log n)(\log\log n)^{1+\delta} \psi_n^2 \sum_{j=n}^{U} c_j^2 < \infty. \tag{2.3.2}$$

Then, there is a random variable S such that

$$\lim_{n\to\infty} \| \sum_{i=1}^{n} Z_i - S\| = 0 \text{ a.s.}$$

If $H=\mathbb{R}$, $\psi_0=1$, $\psi_m=0$ for $m\in N^*$ and $c_n^2 = E\|Z_n\|^2$, then as a particular case we get the stability theorem of martingale differences (Chow 1967). If instead of (2.3.2) we have the condition

$$\sum_{n=1}^{\infty} (\log n)(\log\log n)^{1+\delta} \psi_n^2 < \infty, \tag{2.3.2'}$$

then we get McLeish's theorem. In the case when $H=\mathbb{R}$, theorem 2.3.3. is proved by Masry and Györfi (1987).

Theorem 2.3.3 and the Kronecker Lemma have the following consequence.

__Corollary 2.3.1.__ Let $\{(Z_n, F_n)\}$ be a simple mixingale with parameters $\{c_n\}$ and $\{\psi_m\}$ such that

$$\sum_{j=1}^{\infty} \frac{c_j^2}{j^2} < \infty, \tag{2.3.3}$$

and such that for some $\delta>0$

$$\sum_{n=1}^{\infty} (\log n)(\log\log n)^{1+\delta} \psi_n^2 \sum_{j=n}^{\infty} \frac{c_j^2}{j^2} < \infty. \tag{2.3.4}$$

Then we have

$$\lim_{n\to\infty} \Big\|\frac{1}{n}\sum_{i=1}^{n} Z_i\Big\| = 0 \text{ a.s.}$$

Corollary 2.3.2. Let $(Z_n)_N$ be a sequence of H valued, 0 mean, ρ-mixing random variables for which

$$\sum_{j=1}^{\infty} \frac{E\|Z_j\|^2}{j^2} < \infty, \tag{2.3.5}$$

and such that for some $\delta>0$

$$\sum_{n=1}^{\infty} (\log n)(\log\log n)^{1+\delta} \rho_n^2 \sum_{j=1}^{\infty} \frac{E\|Z_j\|^2}{j^2} < \infty, \tag{2.3.6}$$

Then we have

$$\lim_{n\to\infty} \Big\|\frac{1}{n}\sum_{i=1}^{n} Z_i\Big\| = 0.$$

Proof of corollary 2.3.2. Apply Corollary 2.3.1 for $F_n=F_1^n$ if $n\geq 1$, $F_n = \{\phi,\Omega\}$ if $n\leq 0$, $c_n^2 = E\|Z_n\|^2$ and $\rho_n^2 = \psi_n^2$. We have only to prove c of definition 2.3.2, namely that

$$E \| E(Z_n|F_{n-m}) \|^2 \leq \rho_m^2 \, E \| Z_n \|^2, \tag{2.3.7}$$

which holds because of

$$\begin{aligned} E \| E(Z_n|F_{n-m}) \|^2 &= E \, (E(Z_n|F_{n-m}), E(Z_n|F_{n-m})) \\ &= E \, (Z_n, E(Z_n|F_{n-m})) \\ &\leq \rho_m (E \| Z_n \|^2)^{1/2} \, E \| E \, (Z_n|F_{n-m}) \|^2)^{1/2}. \end{aligned}$$

Corollary 2.3.3. Let $(Z_n)_N$ be a sequence of H valued, 0 mean, α-mixing random variables for which

$$\|Z_j\| \leq c_j \text{ a.s. for } j\in N^*, \tag{2.3.8}$$

where $\{c_j\}$ are some reals such that

$$\sum_{j=1}^{\infty} \frac{c_j^2}{j^2} < \infty.$$

Assume also that for some $\delta>0$ we have

$$\sum_{n=1}^{\infty} (\log n)(\log \log n)^{1+\delta} \alpha_n \sum_{j=n}^{\infty} \frac{c_j^2}{j^2} < \infty.$$

Then we have

$$\lim_{n\to\infty} \| \frac{1}{n} \sum_{j=1}^{n} Z_i \| = 0 \text{ a.s.}$$

<u>**Proof of corollary 2.3.3.**</u> By theorem 2.2.3. we get

$$E \| E(Z_n|F_{n-m}) \|^2 = E\,(Z_n, E(Z_n|F_{n-m})) \le c_n^2 10\, \alpha_m,$$

and we apply Corollary 2.3.1 to obtain the claimed result.

2.4. Dependence characterized by densities.

When we investigate density estimation from dependent samples we need conditions on multidimensional densities. Sometimes these are the only conditions (Rosenblatt, 1971, Castellana and Leadbetter, 1986 and Györfi, 1981), sometimes these are combined with mixing conditions (Masry, 1983 and 1986). We list some of the commonly used conditions.

<u>**Condition D1.**</u> Assume that $(Z_i)_N$ is stationary and ergodic, and there is an integer $m>0$ such that the conditional distribution of Z_m given $F_{-\infty}^{0}$ is absolutely continuous a.s.

A stationary and ergodic Gaussian process satisfies the Condition D1 with m=1. For stationary and ergodic Markov process the Condition D1 means that there is an m such that the m-th transition probability distributions are absolutely continuous. A linear process satisfies Condition D1 with m=1 if Y_0 has a density.

<u>**Conditon D2.**</u> Assume that $(Z_i)_N$ is stationary and there is an integer m such that the density $f_k(x,y)$ of (X_0,X_k) exists for all $k\ge m$.

Another condition we employ is

<u>**Condition D3.**</u> Assume condition D2 and

$$\sup_{x,y\in\mathbb{R}^d} \sum_{k=m}^{\infty} |f_k(x,y)-f(x)f(y)| < \infty,$$

where f stands for the density of Z_1.

Chapter III

REGRESSION ESTIMATION AND TIME SERIES ANALYSIS.

1. Introduction.

Let $(X_i,Y_i)_N$ be a sequence of random pairs valued in $\mathbb{R}^d \times \mathbb{R}$ such that the regression function of Y on X, defined for $x \in \mathbb{R}^d$ by

$$r(x) = E(Y_i | X_i = x) \text{ for any } i=1,2,\ldots, \qquad (3.1.1)$$

exists. Such a condition is in particular satisfied when the process (X_i,Y_i) is stationary, but this assumption will not be necessary for a lot of results presented here. Here r has to be understood as an arbitrary element of the equivalence class of functions defined by (3.1.1). The knowledge of r is helpful in constructing estimates of future values of Y given X)x and it is useful in understanding the relation between the variables X and Y.

A nonparametric estimate of the regression function r is of the form

$$r_n(.) = r_n(.,X_1,Y_1,\ldots,X_n,Y_n),$$

where r_n is a measurable function of its arguments.

In a first course we will look at the universal consistency property in regression estimation (section 2).

The main section of this chapter is section 3 in which we will give uniform convergence results of the more usual nonparametric regression estimate, namely the Watson-Nadaraya kernel estimate, under some mixing conditions.

The introduction of a dependence structure is of particular interest in regression estimation because of the possible applications in Time Series Analysis and Prediction problems. Such problems are discussed in section 4 for processes satisfying mixing conditions. They will be shortly discussed in section 5 for processes which are stationary and ergodic.

In a final section 6, we will give similar results in the setting of estimation of derivatives of r.

This chapter contains only asymptotic consistency results and does not investigate the problem of how choosing the smoothing parameter. This problem is the main difficulty in practical situations and chapter VI of this book is specially devoted to it.

2. Universal consistency in regression estimation.

Let (X,Y) be a pair of random variables such that X takes values in $\mathbb{R}^d$ and Y is real valued and has a finite expectation. The regression function of Y given X is defined by the conditional expectation of Y given X as follows:

$$r(x) = E(Y|X=x), \ x\in\mathbb{R}^d. \qquad (3.2.1)$$

Denote by $D_n=\{(X_1,Y_1),(X_2,Y_2),\dots,(X_n,Y_n)\}$ a stationary sequence of random pairs such that (X,Y) and (X_i,Y_i) have the same distribution, and (X,Y) and D_n are independent one from each other. The sequence D_n is the observed data. A nonparametric estimate of the regression function r is of the form

$$r_n(x) = r_n(x,D_n), \ x\in\mathbb{R}^d, \qquad (3.2.2)$$

where r_n is a measurable function of its arguments.

One might be interested in regression estimates which are consistent for all possible distributions of (X,Y). This generality, unfortunately, produces a lot of mathematical problems. We do not want to escape these mathematical problems, but we can offer only a small part of this book in this full generality. In practice however we believe it is very important to have *some* nonparametric approximation, in particular when we have not enough information for construction of a parametric model. Common assumptions on the unknown distribution of (X,Y) (which may be hard to verify) are the following conditions:

i. r is continuous or k times differentiable,
ii. X has a density.

In the attempt of making convergence results more general Stone (1977) introduced the concept of universal consistency.

Definition 3.2.1. We say that r_n is weak L_p-L_q universal consistent if

$$\lim_{n\to\infty} E\int |r_n(x)-r(x)|^p\mu(dx) = 0$$

for all possible distributions of (X,Y) with $\|Y\|_{L_q}<\infty$, $(q\geq p\geq 1)$, where μ denotes the probability measure of X.

For independent and identically distributed data, Stone (1977) considered a large class of nonparametric regression estimates of the form

$$r_n(x) = \sum_{i=1}^{n} W_{ni}(x,X_1,\dots,X_n)Y_i,$$

and gave the necessary and sufficient condition of their weak consistency in L_p $(p < \infty)$. These results are important tools for verifying the weak L_p-L_p universal consistency of several regression estimates. For the sake of illustration, Stone(1977) applied his results for proving the weak L_p-L_p universal consistency of nearest neighbor estimates.

For kernel estimates Devroye and Wagner (1980) and Spiegelman and Sacks (1980) proved the weak L_p-L_p universal consistency by implicitly verifying Stone's conditions. Unfortunately there are no such general results on strong consistency.

<u>Definition 3.2.2.</u> We say that r_n is strong L_p-L_q universal consistent if

$$\lim_{n\to\infty} \int |r_n(x)-r(x)|^p \mu(dx) = 0 \text{ a.s.}$$

for all possible distribution of (X,Y) with $\|Y\|_{L_q} < \infty$ for $q \geq p \geq 1$.

Devroye and Krzyzak (1987) proved that for the kernel estimate the weak and the strong L_1-L_∞ universal consistency properties are equivalent. There are concepts for pointwise universal consistencies. In this respect the kernel estimate was considered by Devroye (1981) and Györfi (1981b), the recursive kernel estimate by Greblicki, Krzyzak and Pawlak (1984), Krzyzak and Pawlak (1983, 1984a and 1984b), Greblicki and Pawlak (1987), and the nearest neighbor estimate by Devroye (1981), Devroye and Wise (1980).

In the sequel we prove the strong L_1-L_∞ universal consistency of partitioning estimates which is defined as follows: let $P_n=\{A_{n1}, A_{n2},\dots\}$ be a sequence of partitions such that $0<\lambda(A_{ij})$, $i=1,2,\dots$, $j=1,2,\dots$, where λ stands for the Lebesgue measure. If μ_n denotes the empirical measure for the sample $\{X_1,X_2,\dots X_n\}$ and for a set A

$$\nu_n(A) = \frac{1}{n}\sum_{i=1}^{n} I_{\{X_i \in A\}} Y_i, \tag{3.2.3}$$

then the partitioning estimate of the regression function r is defined by

$$r_n(x)=\begin{cases} \dfrac{\nu_n(A_{ni})}{\mu_n(A_{ni})}, & \text{if } x\in A_{ni} \text{ and } \mu_n(A_{ni}) > 0, \\ \dfrac{1}{n}\displaystyle\sum_{i=1}^{n} Y_i, & \text{if } x\in A_{ni} \text{ and } \mu_n(A_{ni}) = 0. \end{cases} \tag{3.2.4}$$

For independent sample we have the following (Devroye and Györfi, 1985b): assume that $|Y| \leq M < \infty$ a.s., where M is a constant and for each sphere S

$$\lim_{n\to\infty} \frac{1}{n} \#\{i;\ A_{ni}\cap S\neq\Phi\}^{\delta} = 0, \tag{3.2.5}$$

where $\delta=1$ and

$$\lim_{n\to\infty} \max_{i:A_{ni}\cap S\neq 0} \{ \sup_{x,y\in A_{ni}} \|x-y\|_{R^d} \} = 0. \tag{3.2.6}$$

Then for the partitioning estimate (3.2.4) and for each $\epsilon>0$ there exists n_o such that

$$P(\int |r_n(x)-r(x)|\mu(dx)\geq\epsilon) < e^{-cn(\frac{\epsilon}{M})^2}, \text{ for } n>n_o, \tag{3.2.7}$$

where c is a universal constant.

Obviously (3.2.7) implies that under the conditions (3.2.5) and (3.2.6) the partioning estimate is strong L_1-L_∞ universal consistent.

We say that P_n is a cubic partition if A_{ni} is a rectangle of form

$$\prod_{j=1}^{d} [a_j k_{ij} h_n,\ a_j(k_{ij}+1)h_n),$$

where $a_1, a_2, \ldots a_d$ are real and $\{k_{ij}\}$ are integers. Then, (3.2.6) is equivalent to

$$\lim_{n\to\infty} h_n = 0,$$

while (3.2.5) is equivalent to

$$\lim_{n\to\infty} nh_n^{\delta d} = \infty.$$

<u>**Theorem 3.2.1.**</u> Assume that (3.2.6) holds. Then any of the following two assumptions implies the strong L_1-L_∞ universal consistency of partitioning estimate (3.2.4):

i. $(X_i)_N$ is φ-mixing and (3.2.5) is met with $\delta=1$;

ii. $(X_i)_N$ is α-mixing and for a $\delta_1>0$

$$\sum_{i=1}^{\infty} i^{\delta_1}\alpha_i < \infty, \tag{3.2.8}$$

moreover for $\delta=\delta_2>2\frac{1+\delta_1}{\delta_1}$ assume (3.2.5).

In the proof of (3.2.7) the following lemmas plaid a key role. The proof of theorem 3.2.1 will be given after the statement of these lemmas.

<u>**Lemma 3.2.1.**</u> (Devroye and Györfi, 1985a, lemma 3.1). If $P_n=\{A_{n1},\dots A_{nk}\}$ is a partition of $\mathbb{R}^d$ and $0<\epsilon<1$ then for $k/n<\epsilon^2/20$ we have

$$P\left(\sum_{i=1}^{k}|\mu_n(A_{ni})-\mu(A_{ni})|>\epsilon\right) \le \exp(-n\epsilon^2/25). \qquad (3.2.9)$$

Let us note that Devroye (1987) extended (3.2.9) in an elegant way as follows,

$$P\left(\sum_{i=1}^{k}|\mu_n(A_{ni})-\mu(A_{ni})|>\epsilon\right) \le 2^{k+1}\exp(-n\epsilon^2/2)$$

So, we have to develop lemmas for dependent samples similar to lemma 3.2.1.

<u>**Lemma 3.2.2.**</u> If $P_n=\{A_{n1},\dots A_{nk}\}$ is a partition, and $(X_i)_N$ is stationary and φ-mixing, then for each $\epsilon>0$ there is m_0 depending on $(\varphi_i)_N$ and $\epsilon>0$ such that for $k/n<\epsilon/32m_0$ we have

$$P\left(\sum_{i=1}^{k}|\mu_n(A_{ni})-\mu(A_{ni})|>\epsilon\right) \le \exp\left(-n\frac{\epsilon}{32m_0}\right). \qquad (3.2.10)$$

<u>**Proof of lemma 3.2.2.**</u> Apply theorem 2.2.1. for

$$Z_i=\frac{1}{n}\left(I_{\{X_i\in A\}}-\mu(A)\right),\ \alpha=\frac{n}{4m},\ d=\frac{1}{n},\ \delta=\frac{1}{2n}\text{ and } D=\frac{1}{4n^2}$$

then

$$P(|\mu_n(A)-\mu(A)|>\epsilon) \le \exp\left(3\sqrt{e}\,\frac{\varphi_m}{m}-\frac{\epsilon n}{4m}+\frac{3\left(1+8\sum_{i=1}^{m}\varphi_i\right)}{32m^2}n\right).$$

Choose m_0 such that

$$\frac{3\left(1+8\sum_{i=1}^{m_0}\varphi_i\right)}{32m_0}<\frac{\epsilon}{16}\text{ and } 3\sqrt{e}\,\varphi_{m_0}<\frac{\epsilon}{16},$$

then

$$P(|\mu_n(A)-\mu(A)|>\epsilon) \le \exp(-\frac{n\epsilon}{8m_0}). \qquad (3.2.11)$$

Since μ_n and μ are probability measures and P_n is a partition, we have

$$\sum_{i=1}^{k} |\mu_n(A_{ni}) - \mu(A_{ni})| = 2 \sup_{A\in F(P_n)} (\mu_n(A) - \mu(A)).$$

Applying (3.2.11) we get

$$P(\sum_{i=1}^{k} |\mu_n(A_{ni}) - \mu(A_{ni})|>\epsilon) \le 2^k \sup_{A\in F(P_n)} P(2|\mu_n(A) - \mu(A)|>\epsilon)$$
$$\le e^{k\log 2-n\epsilon/(16m_0)} \le e^{-n\epsilon/(32m_0)}.$$

This completes the proof of lemma 3.2.2.

Applying a slight modification of the proof of lemma 1 in Devroye and Györfi (1985b) (using lemma 3.2.2. instead of Lemma 3.2.1) we get the following lemma.

<u>Lemma 3.2.3.</u> If $P_n=\{A_{n1},\dots A_{nk}\}$ is a partition, and $(X_i,Y_i)_N$ is stationary and φ-mixing, then for each $\epsilon>0$ there is m_0 depending on $(\varphi_i)_N$, M and $\epsilon>0$ such that for $k/n < \epsilon/(32m_0)$ we have

$$P(\sum_{i=1}^{k} |\nu_n(A_{ni})-\nu(A_{ni})|>\epsilon) \le \exp(-n\frac{\epsilon}{32m_0}). \qquad (3.2.12)$$

<u>Lemma 3.2.4.</u> If $P_n=\{A_{n1},A_{n2},\dots,A_{nk_n}\}$, $n=1,2,\dots$ is a sequence of partitions and $(X_i,Y_i)_N$ is stationary and α-mixing such that

$$\sum_{i=1}^{\infty} i^{\delta_1}\alpha_i < \infty \text{ for some } \delta_1 > 0, \qquad (3.2.13)$$

and $k_n^{\delta_2}/n$ is bounded for

$$\delta_2 > 2\frac{1+\delta_1}{\delta_1}, \qquad (3.2.14)$$

then there exists $\delta_3 > 0$ such that for each $\epsilon>0$ there is a constant c for which

$$P\left(\sum_{i=1}^{k_n} |\mu_n(A_{ni}) - \mu(A_{ni})| > \epsilon\right) \le cn^{-1-\delta_3}. \tag{3.2.15}$$

<u>**Proof of lemma 3.2.4.**</u> For a Borel set A put

$$Z_i = I_{\{X_i \in A\}} - \mu(A)$$

and apply theorem 2.2.4. Then we have for some finite real positive constant C

$$E|n(\mu_n(A)-\mu(A))|^{2+2\delta_1} \le C\, n^{1+\delta_1},$$

and therefore

$$P\left(\sum_{i=1}^{k_n} |\mu(A_{ni}) - \mu_n(A_{ni})| > \epsilon\right)$$

$$\le \frac{k_n^{2+2\delta_1-1} \sum_{i=1}^{k_n} E|n(\mu(A_{ni}) - \mu_n(A_{ni}))|^{2+2\delta_1}}{n^{2+2\delta_1}\ \epsilon^{2+2\delta_1}}$$

$$\le \frac{k_n^{2+2\delta_1}\ C\, n^{1+\delta_1}}{n^{2(1+\delta_1)}\ \epsilon^{2+2\delta_1}} = \frac{k_n^{2+2\delta_1}\ C}{n^{1+\delta_1}\ \epsilon^{2+2\delta_1}}. \tag{3.2.16}$$

For given δ_1 and δ_2 let δ_3 be the solution of $\dfrac{2+2\delta_1}{\delta_1 - \delta_3} = \delta_2$ which exists by (3.2.14). Then we have

$$P\left(\sum_{i=1}^{k_n} |\mu(A_{ni}) - \mu_n(A_{ni})| > \epsilon\right)$$

$$\le \frac{C}{\epsilon^{2+2\delta_1}} \left(\frac{k_n^{\delta_2}}{n}\right)^{\delta_1-\delta_3} \frac{1}{n^{1+\delta_3}}, \tag{3.2.17}$$

and the lemma is proved.

Applying again the modification of the proof of Lemma 1 in Devroye and Györfi (1985b) (using lemma 3.2.4. instead of lemma 3.2.1) we get the following lemma.

<u>**Lemma 3.2.5.**</u> Under the conditions of lemma 3.2.4, there exists $\delta_3 > 0$ such that for each $\epsilon > 0$ there is a constant c for which

$$P\left(\sum_{i=1}^{k_n} |\upsilon_n(A_{ni}) - \upsilon(A_{ni})| > \epsilon\right) \leq \frac{c}{n^{1+\delta_3}}. \tag{3.2.18}$$

<u>**Lemma 3.2.6.**</u> (Devroye and Györfi, 1985b, lemma 2). Let S be a sphere and put

$$I_{Sn} = \{i; A_{ni} \cap S \neq 0\},$$

then under the condition (3.2.6) we get

$$\lim_{n\to\infty} \sum_{i\in I_{Sn}} \int_{A_{ni}} \left|\frac{\upsilon(A_{ni})}{\mu(A_{ni})} - r(x)\right| \mu(dx) = 0.$$

<u>**Proof of theorem 3.2.1.**</u> Let C be the bound on $|Y|$ thus $|Y| \leq C$ a.s. then for each sphere S it is easy to verify (see the proof of theorem in Devroye and Györfi, 1985b) that

$$\begin{aligned}\int |r_n(x) - r(x)| \mu(dx) \leq\ & 2C\mu(S^c) + C \sum_{i\in I_{Sn}} |\mu_n(A_{ni}) - \mu(A_{ni})| \\ & + \sum_{i\in I_{Sn}} |\upsilon_n(A_{ni}) - \upsilon(A_{ni})| \\ & + \sum_{i\in I_{Sn}} \int_{A_{ni}} \left|\frac{\upsilon(A_{ni})}{\mu(A_{ni})} - r(x)\right| \mu(dx).\end{aligned} \tag{3.2.19}$$

By lemma 3.2.6 the last term of (3.2.19) converges to 0.

<u>**Proof of i.**</u> Let k_n be the size of the set I_{Sn} then the second and the third terms of the right hand side of (3.2.19) tend to 0 because of lemmas 3.2.2 and 3.2.3., resp. since condition (3.2.5) implies that $k_n/n \to 0$.

<u>**Proof of ii.**</u> Let k_n be as before. Then (3.2.5) implies that $k_n^{\delta}/n \to 0$ where $\delta = \delta_2$. Therefore lemmas 3.2.4. and 3.2.5. imply the convergence of the second and third terms of the right hand side of (3.2.19).

<u>3. Uniform convergence of kernel estimates under mixing conditions.</u>

<u>3.1. Kernel estimate under independence.</u>

Many authors have investigated non parametric techniques to estimate the function r when the random variables $(X_i, Y_i)_N$ are independent (see the surveys by Collomb, 1981 and 1985a). The most frequently analyzed non parametric regression estimate (Watson, 1964 and Nadaraya, 1964) is defined from a kernel function K of $\mathbb{R}^d$ and a sequence of smoothing parameters $(h_n)_N$ by

$$r_n(x) = \sum_i Y_i K((x-X_i)/h_n) \;/\; \sum_i K((x-X_i)/h_n). \qquad (3.2.1)$$

The parameter h_n, which controls the degree of smoothness of the estimator r_n, is called the bandwidth. We will abreviate it from now on to h. In the following we will only consider this class of kernel estimates. As pointed out in the surveys already cited, most of other non parametric techniques can be seen as extensions or as specializations of these Watson-Nadaraya estimators, see Härdle (1989). The main reason to limit oneself to kernel estimates is that they reach the optimal rates of convergence over a large class of regression functions when the observations (X_i,Y_i) are independent. This is precised through the following result. Under appropriate conditions on K, h and on the probability law of (X,Y) we have the following result.

Theorem 3.3.0. For independent pairs $(X_i,Y_i)_N$ and if r is k-times differentiable and if its derivatives of order k are Lipschitz continuous with order γ, then we have for any compact subset G of $\mathbb{R}^d$

$$\sup_{x \in G} |r_n(x)-r(x)| = O\,[(n^{-1}h^{-d}\log n)^{1/2}] + O[h^{2(k+\gamma)}], \text{ co.}$$

Here we used the symbol "co" to denote complete convergence.

Definition 3.3.1. A sequence $(Z_i)_N$ of random variables *converges completely* to 0 (abbreviated in $Z_i \xrightarrow{\text{co.}} 0$) if there exists some positive real a such that we have

$$\sum_{i=1}^{\infty} P(Z_i > a) < \infty.$$

Let us also note that (via Borel-Cantelli lemma) such a mode of convergence implies as well convergence in probability as almost sure convergence.

The rate given in theorem 3.1.0 above can be optimised by taking h proportional to $(n^{-1}\log n)^{1/(2k+2\gamma+d)}$, leading to a rate of convergence in $(n^{-1}\log n)^{(k+\gamma)/(2k+2\gamma+d)}$, which has been shown (Stone (1982)) to be the optimal global rate of convergence for regression function of smoothness $k+\gamma$.

The aim of this section is to give similar results under several mixing conditions. It is necessary to note that, for independent data, there are two ways to prove consistency results. The first route consists in using strong approximations of empirical process and is often limited (beside independence) to the case d=1 (see Mack and Silverman, 1982 for a proof of theorem 3.3.0 by this way). It seems very difficult to apply these techniques to the dependent and multivariate case. We opted for the second route which consists mainly in applying analytic techniques, (see Härdle and Luckaus, 1984 for a proof of theorem 3.3.0 by this way and also Collomb, Härdle and Hassani, 1986). These techniques are adapted to the dependent case by using exponential type inequalities following the previous work of Collomb (1984). We note that theorem 3.3.0 above is a trivial consequence of theorem 3.3.2 below.

3.2. General assumptions.

Let us denote by G a compact subset of $\mathbb{R}^d$, and by $\hat{G}$ an ϵ-neighborhood compact of G ($G\subset\hat{G}$). In order to insure the uniform consistency of r_n to r on G we need the following assumptions on the probability distribution of (X,Y).

$$\exists\Gamma<\infty,\ \forall i\in N,\ \forall B\in B(\mathbb{R}^d)\ P(X_i\in B) \leq \Gamma\lambda(B),$$

and (A.1)

$$\exists\gamma\in>0,\ \forall i\in N,\ \forall B\in B(\hat{G})\ P(X_i\in B) \geq \gamma\lambda(B),$$

where $B(\mathbb{R}^d)$ (resp. $B(\hat{G})$) is the σ-algebra of the Borel sets on $\mathbb{R}^d$ (resp. on $\hat{G}$), and λ the Lebesgue measure on $\mathbb{R}^d$. We assume the existence of absolute moments for the random variables Y_i, i.e.,

$$\exists\beta>2,\ \exists M<\infty,\ \forall i\in N\ E(|Y_i|^\beta) \leq M. \tag{A.2}$$

The conditional variances are assumed to be bounded on $\hat{G}$, i.e.,

$$\exists V<\infty,\ \forall i\in N,\ \forall x\in\hat{G}\ E[((Y_i-r(X_i))^2|X_i=x] \leq V. \tag{A.3}$$

To specify the rate of convergence of r_n we will need to assume the existence of a common marginal density f of the variables X_i, and some smoothness assumptions on the functions r and f, namely that for some $k\in N$ and some $\gamma\in(0,1)$

r and f are k-times differentiable
and their derivatives of order k (A.4)
are Lipschitz continuous of order γ.

Remark 3.3.1. The condition (A.1) is often used in kernel regression estimation to insure that the denominator of r_n does not vanish on G; it is in particular satisfied when the variables X_i have a common continuous marginal density which is bounded away from zero on G.
The conditions (A.2) and (A.3) may appear somewhat restrictive, but it should be noted that most papers dealing with kernel estimators, even in the independent case, introduce stronger conditions like assuming that the

variables Y_i are uniformly bounded (see the surveys by Collomb, 1981 and 1985a).
Note also that k in (A.4) is allowed to be 0.

The kernel function K will be supposed to satisfy the following conditions:

$$\exists \bar{K},\ \forall x \in \mathbb{R}^d,\ |K(x)| \leq \bar{K} < \infty, \qquad (K.1)$$

$$\|x\|^d K(x) \longrightarrow 0 \text{ as } \|x\| \longrightarrow \infty, \qquad (K.2)$$

$$\exists \hat{K},\ |\int K(u)du| \leq \hat{K} < \infty, \qquad (K.3)$$

$$K \text{ is Lipschitz continuous of order } \gamma \text{ on } \mathbb{R}^d, \qquad (K.4)$$

$$\int \|u\|^j K(u)du = 0 \text{ for } j=1,\dots,k, \qquad (K.5)$$

$$\exists K^*,\ 0 < |\int \|u\|^{k+1} K(u)du| \leq K^* < \infty. \qquad (K.6)$$

<u>**Remark 3.3.2.**</u> The conditions (K.1), (K.2) and (K.3) are quite usual in kernel estimation. The condition (K.4) is only introduced to make the proofs clearer; the reader will find in Collomb, Hassani, Sarda and Vieu (1985) how by applications of basic topological results we can extend the properties obtained for Lipschitz continuous kernels to discontinuous ones. The ideas of the proof for such an extension are the same as those described in the setting of hazard estimation in remark 5.4.2 below. The conditions (K.5) and (K.6) will be used only to specify the rate of convergence.

3.3. Case of φ-mixing variables.

Let us assume through the following section that the random variables $(X_i,Y_i)_N$ are φ-mixing, following the definition 2.2.1. The properties of the estimator r_n will depend on the mixing coefficients $(\varphi_n)_N$ through an increasing sequence $(m_n)_N$ of integers such that

$$\exists A<\infty,\ \forall n \in N,\ 1 \leq m_n \leq n \text{ and } n\varphi_{m_n}/m_n \leq A. \qquad (3.3.1)$$

<u>**Remark 3.3.3.**</u> We note that when the process $(X_i,Y_i)_N$ has geometrically decreasing mixing coefficients (φ_n), the sequence (m_n) could be taken proportional to logn. Examples of such a case have been given in section 2.4. of chapter II. In the case of independent observations, m_n could be taken to be 1 for any n.

In order to deal with possible high values for the variables Y_i, let us denote by M_n an increasing sequence satisfying

$$M_n = n^{\xi}, \text{ for some } \xi \in (4(\beta+2)^{-1}, 1). \tag{3.2.2}$$

Theorem 3.3.1. Uniform convergence. (Sarda and Vieu, 1985a). Assume that the conditions (A.1)–(A.3) and (K.1)–(K.4) hold. If the function r is continuous on G and if the bandwidth h is such that

$$nh^d/(m_n M_n \log n) \longrightarrow \infty, \tag{H.1}$$

then we have

$$\sup_{x \in G} |r_n(x) - r(x)| \xrightarrow{co.} 0.$$

The proof of this theorem is not given since it follows obviously from theorem 3.3.2 below. In the particular case when the variables Y_i are assumed to be uniformly bounded, this result has been given by Collomb (1984), with the change that $M_n = 1$, $\forall n \in N$.

The following result specifies the rate of convergence of the estimate r_n as a function of the smoothness of the regression function r.

Theorem 3.3.2. Rate of convergence. Assume that conditions (A.1)–(A.4) and (K.1)–(K.6) hold. If the bandwidth is such that the sequence

$$V_n = h^{k+\gamma} + [(m_n M_n \log n)/(nh^d)]^{1/2},$$

satisfies the condition

$$V_n \longrightarrow 0 \text{ as } n \longrightarrow \infty, \tag{H.2}$$

then we have

$$\sup_{x \in G} |r_n(x) - r(x)| = O(V_n), \text{ co.}$$

Before giving the proof of this theorem, let us mention some additional results. (These remarks will be proven just after the proof of theorem 3.3.2.)

Remark 3.3.4. If in place of (A.2) we assume

$$\exists M<\infty, \ \exists \alpha>0, \ \forall i \in N \quad E\exp(\alpha Y_i) \leq M, \tag{A.2.a}$$

we get a rate of convergence in

$$V_n = h^{k+\gamma} + [(m_n \log^2 n)/(nh^d)]^{1/2}.$$

Remark 3.3.5. If in place of (A.2) we assume

$$\exists M<\infty,\ \forall i\in N\quad |Y_i| \le M, \qquad \text{(A.2.b)}$$

we get the rate of convergence

$$V_n = h^{k+\gamma} + [(m_n \log n)/(nh^d)]^{1/2}.$$

This result was given by Collomb and HCrdle (1986) in the particular case when k=2 and γ=0.

These rates of convergence can be optimised by a suitable asymptotic choice of the bandwidth h.

Corollary 3.3.1. Assume that the conditions of theorem 3.3.2 hold and that the bandwidth is taken to be such that for some finite positive constant C

$$h^* = C\ (n^{-1} m_n M_n \log n)^{1/(2k+2\gamma+d)}, \qquad (3.3.3)$$

then we obtain a rate of convergence in

$$V_n^* = (n^{-1} m_n M_n \log n)^{(k+\gamma)/(2k+2\gamma+d)}.$$

Proof of corollary 3.3.1. It suffices to note that h^* defined in (3.3.3) balances the trade-off between the two components of V_n.

Proof of theorem 3.3.2. Let us express the estimate r_n in the following form

$$r_n = g_n/f_n,$$

with

$$g_n(x) = (nh^d)^{-1} \sum_i Y_i K((x-X_i)/h),$$

and

$$f_n(x) = (nh^d)^{-1} \sum_i K((x-X_i)/h).$$

We have the following decomposition

$$r_n - r = (A_1 + A_2 + rA_3 + rA_4) f_n^{-1},$$

where

$$A_1 = g_n - Eg_n,$$

$$A_2 = Eg_n - g,$$

$$A_3 = f_n - Ef_n,$$

$$A_4 = Ef_n - f.$$

The function r is bounded on G because of (A.4) and the kernel estimate f_n of f is almost surely bounded away from zero on G because of (A.1) and of 3.3.4.b and 3.3.4.d below. Therefore it is enough to show that

$$\sup_{x \in G} A_1(x) = O((n^{-1}h^{-d}m_nM_n\log n)^{1/2})), \qquad (3.3.4)$$

$$\sup_{x \in G} A_2(x) = O(h^{k+\gamma}), \qquad (3.3.4a)$$

$$\sup_{x \in G} A_3(x) = O((n^{-1}h^{-d}m_nM_n\log n)^{1/2}), \qquad (3.3.4b)$$

$$\sup_{x \in G} A_4(x) = O(h^{k+\gamma}). \qquad (3.3.4c)$$

For which concerns the bias terms A_2 and A_4, they can be treated exactly as in the independent case since these terms does not depend on the probability distribution of $(X_1,Y_1),\ldots,(X_n,Y_n)$. Therefore the proof of (3.3.4.a and c) is omitted, it can be found in Härdle and Luckhaus (1984). We just mention that it is based on Taylor expansions of the functions f and g and on the utilisation of the classical so-called Bochner theorem (see e.g. Parzen, 1962 and Collomb, 1976 and 1984). We mention also in this topic the related works of Lejeune (1985) and the references included in it.

Clearly, the term A_3 can be seen as a particular case of the term A_1 (in which $Y_i=1$, $\forall i \in N$). Therefore the proof will be complete by showing (3.3.4).

The study of g_n can not be made directly because of the possible large values for the variables Y_i. Following the idea of Mack and Silverman (1982) a truncation technique is used (see also Sarda and Vieu, 1985a and 1985b). This consists in decomposing g_n in g_n^+ and g_n^- where

$$g_n^+(x) = (nh^d)^{-1} \sum_i Y_iK((x-X_i)/h)\, I_{\{|Y_i| \geq M_n\}},$$

and

$$g_n^-(x) = g_n(x) - g_n^+(x). \qquad (3.3.5)$$

Using the condition (K.1) we have

$$\sup_{x\in\mathbb{R}^d} |g_n^+(x)| \le \hat{K}(nh^d)^{-1} \sum_i |Y_i| \, I_{\{|Y_i|\ge M_n\}},$$

leading by Schwartz inequality to

$$E \sup_{x\in\mathbb{R}^d} |g_n^+(x)| \le \hat{K}(nh^d)^{-1} \sum_i (EY_i^2)^{1/2}(P(|Y_i|\ge M_n))^{1/2}.$$

Using now Chebyshev's inequality and (A.2) we get for any sequence $(\epsilon_n)_N$

$$P(\sup_{x\in\mathbb{R}^d} |g_n^+(x)-Eg_n^+(x)|\ge\epsilon_n) \le 2\epsilon_n^{-1}\hat{K}h^{-d}M^{1/2}(M/M_n^{\beta})^{1/2}$$
$$= O(\epsilon_n^{-1}h^{-d}M_n^{-\beta/2}).$$

It suffices to choose now $\epsilon_n=\epsilon_0(nh^d)^{-1}$ and to use the definition of M_n given in (3.3.2) to get

$$P(M_n^{-1}nh^d \sup_{x\in\mathbb{R}^d} |g_n^+(x)-Eg_n^+(x)|\ge\epsilon_0) \le LnM_n^{-1-\beta/2} < \infty. \quad (3.3.6)$$

Then (3.3.4) will follow from (3.3.5), (3.3.6) and the following lemma.

<u>**Lemma 3.3.1.**</u> Under the conditions of theorem 3.3.2 we have for any $\epsilon>0$,

$$\sum_{n=1}^{\infty} P(V_n \sup_{x\in G} |g_n^-(x) - Eg_n^-(x)| > \epsilon) < \infty.$$

<u>**Proof of lemma 3.3.1.**</u> Let us denote along this proof by C a generic constant. The main tool of this proof is the application of Collomb's inequality (Collomb, 1984, lemma 1) which is recalled in theorem 2.2.1.

Let us denote by

$$\Psi_i = n^{-1}h^{-d}Y_i I_{\{|Y_i|\le M_n\}},$$

and

$$Z_i = \Psi_i - E\Psi_i.$$

We have by (K.1),

$$|Z_i| \le n^{-1}h^{-d}M_n\bar{K}.$$

We have also

$$E|Z_i|\le 2n^{-1}E(|Y_i|h^{-d}K((x-X_i)/h)$$

$$\leq Cn^{-1}\int E(|Y_i||X_i=u)h^{-d}K((x-u)/h)du,$$

leading by Schwarz's inequality, (A.3) and then (K.3) to

$$E|Z_i| \leq Cn^{-1}\int(r^2(u)+V)^{1/2}h^{-d}K((x-u)/h)du \leq C\hat{K}n^{-1}.$$

By the same arguments we can show that

$$EZ_i^2 \leq Cn^{-2}h^{-d}\int(r^2(u)+V)h^{-d}K((x-u)/h)du \leq C\hat{K}n^{-2}h^{-d}. \quad (3.3.7)$$

Let us take $m=m_n$ (resp. $m=m_0$) if $\lim m_n = \infty$ (resp. if $\exists n_0, \forall n>n_0\ m_n=m_0$), and choose α such that

$$\alpha = Cnh^d/(mM_n) \text{ and } (\alpha mn^{-1}h^{-d}M_n\bar{K}) > 1/4.$$

Collomb's inequality (Theorem 2.2.1) gives then for any sequence $(\epsilon_n)_N$

$$P(|g_n^-(x)-Eg_n^-(x)|>\epsilon_n) \leq C_1 \exp(-C_2nh^d\epsilon_n^2/(m_nM_n)),$$

for some constant C_1 and C_2 independent of x. That leads to

$$\sup_{x\in\mathbb{R}^d} P(|g_n^-(x)-Eg_n^-(x)|>\epsilon_n)\leq C_1\exp(-C_2nh^d\epsilon_n^2/(m_nM_n)). \quad (3.3.8)$$

Using now the compactness property of G, we cover it by a finite number of balls B_k of center t_k in such a way that

$$\left.\begin{array}{l} G \subset \bigcup_{i=1}^{l_n} B_k, \\ \sup_{x\in B_j} \|x-t_j\| \leq h^\delta,\ \forall j=1,\dots,l_n, \\ l_n \leq Ch^{-\delta d}, \\ \delta>(k+d+2\gamma)\gamma^{-1}+q\gamma^{-1} \text{ for some } q>0. \end{array}\right| \quad (3.3.9)$$

We have

$$\sup_{x\in G}|g_n^-(x)-Eg_n^-(x)| \leq \max_{k=1,\dots l_n}|S_n(t_k)|+\sup_{x\in G}|\hat{S}_n(x)|, \quad (3.3.10)$$

where

$$S_n(x) = g_n^-(x) - Eg_n^-(x)$$

and

$$\hat{S}_n(x) = S_n(x) - S_n(t_{k(x)}),$$

$k(x)$ being the index of the ball $B_{k(x)}$ containing x. By (3.3.8) we have

$$P(\max_{k=1,\dots l_n} |S_n(t_k)|>\epsilon_n) \le l_n \sup_{x\in G} P(|S_n(x)|>\epsilon_n)$$

$$\le C_1 l_n \exp(-C_2 nh^d \epsilon_n^2/(m_n M_n)),$$

which leads with (3.3.9) to

$$\sum_{n=1}^{\infty} P((nh^d)^{1/2}(m_n M_n \log n)^{-1/2} \max_{k=1,\dots,l_n} |S_n(t_k)|>\epsilon_0)<\infty. \quad (3.3.11)$$

We now consider the second term in the decomposition (3.3.10). We have

$$|\hat{S}_n(x)| \le n^{-1} \sum_i h^{-d}|K((x-X_i)/h)-K((t_{k(x)}-X_i)/h)||Y_i|$$

$$+ Cn^{-1} \sum_i E(h^{-d}|K((x-X_i)/h)-K((t_{k(x)}-X_i)/h)||Y_i|).$$

The Lipschitz condition (K.4) can be written as

$$|h^{-d}K(u/h)-h^{-d}K(v/h)| \le Ch^{-(d+\gamma)}|u-v|^{\gamma},\ \forall u,v\in\mathbb{R}^d,$$

so we have

$$\sup_{x\in G} |\hat{S}_n(x)| \le Ch^{-(d+\gamma)}|x-t_{k(x)}|^{\gamma}(n^{-1}\sum_i |Y_i|+E|Y_i|).$$

Since, by definition, x belongs to $B_{k(x)}$ we have by (3.3.9)

$$\sup_{x\in G} |\hat{S}_n(x)| \le Ch^{k+\gamma+q}(n^{-1}\sum_i |Y_i|+E|Y_i|),$$

and because

$$n^{-1} \sum_i |Y_i| \xrightarrow{a.s} E|Y_i| \le (M+1)^{1/2},$$

we get

$$\sup_{x\in G} |\hat{S}_n(x)| \le Ch^{k+q+\gamma},\ \text{a.s.}$$

Since $V_n \longrightarrow 0$, we can find q large enough to have $h^q=O(n^{-(1+\alpha)})$ for some

$\alpha>0$ ans so by Chebyshev's inequality we get

$$\sum_{n=1}^{\infty} P(h^{k+\gamma} \sup_{x\in G}|\hat{S}_n(x)| > \epsilon_0) < \infty. \qquad (3.3.12)$$

The proof of lemma 3.3.1 follows now from (3.3.10), (3.3.11) and (3.3.12). The proof of theorem 3.3.2 is therefore complete.

<u>**Proof of remark 3.3.4.**</u> It suffices to note that under (A.2.a) we can choose M_n=Clogn (in place of (3.3.2)). We still get (3.3.6) for g_n^+ with this new value of M_n, and the truncated term g_n^- is not affected by this change. Details for this can be found in Sarda and Vieu (1985a) and (1985b).

<u>**Proof of remark 3.3.5.**</u> It suffices here to take M_n=M. The term g_n^- does not exist in this case, and g_n^+ is not affected by this change.

3.4. Case of ρ-mixing variables.

Let us assume through the following section that the random variables $(X_i,Y_i)_N$ are ρ-mixing, following the definition 2.2.2. The properties of the estimator r_n will depend on the mixing coefficients through sequences $(d_n)_N$ and $(q_n)_N$ defined by

$$d_n = 8000 \exp(3 \sum_{i=1}^{[\log n]} \rho_n^*(2^i)), \qquad (3.3.13)$$

with

$$\rho_n^*(j) = \max_{1\leq k\leq n-j-1}\ \max_{1\leq m\leq (n-k-j)/2} |E< \sum_{i=k}^{k+m} X_i,\ \sum_{i=k+m+j}^{k+2m+j} X_i>|,$$

and

$$n/q_n \geq 2 \text{ and } \limsup\, [2nq_n^{-1}]^{1/2}\rho(q_n) < 4/9. \qquad (3.3.14)$$

Let us recall that the truncation sequence M_n is an increasing sequence satisfying

$$M_n = n^{\xi}, \text{ for some } \xi \in (4(\beta+2)^{-1},1).$$

<u>**Theorem 3.3.3.**</u> **Uniform convergence.** Assume that the conditions (A.1)-(A.3) and (K.1)-(K.4) hold. If the function r is continuous on G and if the bandwidth h is such that

$$nh^d/(d_n M_n \log^2 n) \longrightarrow \infty$$

and (H.3)

$$nh^d/(q_n \log n) \longrightarrow \infty,$$

then we have

$$\sup_{x \in G} |r_n(x) - r(x)| \xrightarrow{co.} 0.$$

We do not give the proof of this theorem since it is an obvious consequence of theorem 3.3.4 below. This result was proven in Peligrad (1987) for uniformly bounded variables Y_i, with the change that $M_n=1$, $\forall n \in N$.

The following result specifies the rate of convergence of the estimate r_n as a function of the smoothness of the regression function r.

<u>**Theorem 3.3.4.**</u> **Rate of convergence.** Assume that conditions (A.1)-(A.4) and (K.1)-(K.6) hold. If the bandwidth is such that the sequence

$$W_n = h^{k+\gamma} + [(d_n M_n \log^2 n)/(nh^d)]^{1/2} + (q_n \log n)/(nh^d),$$

satisfies the condition

$$W_n \longrightarrow 0 \text{ as } n \longrightarrow \infty, \qquad \text{(H.4)}$$

then we have

$$\sup_{x \in G} |r_n(x) - r(x)| = O(W_n), \text{ co.}$$

Before giving the proof of this result let us make some remarks and derive some corollaries. (These remarks will be proven just after the proof of theorem 3.3.4.)

<u>**Remark 3.3.6.**</u> If in place of (A.2) we assume (A.2.a), we get a rate of convergence in

$$W_n = h^{k+\gamma} + [(d_n \log^3 n)/(nh^d)]^{1/2} + (q_n \log n)/(nh^d).$$

<u>**Remark 3.3.7.**</u> If in place of (A.2) we assume (A.2.b), we get a rate of convergence in

$$W_n = h^{k+\gamma} + [(d_n \log^2 n)/(nh^d)]^{1/2} + (q_n \log n)/(nh^d).$$

It is not possible to optimise the rate of convergence in the general case, as for φ-mixing variables, since W_n depends on two sequences $(d_n)_N$ and $(q_n)_N$. However we can do this in two particular examples, which are the most important in practice.

Corollary 3.3.2. Assume that the conditions of theorem 3.3.4 hold and that we have $\rho_n=O(a^{-n})$ for some a>1. If the bandwidth is taken to be, for some real positive constant C, of the form

$$h^* = C\,(n^{-1}\log^2 n)^{1/(2k+2\gamma+d)}, \tag{3.3.15}$$

then we obtain a rate of convergence in

$$W_n^* = (n^{-1}\log^2 n)^{(k+\delta)/(2k+2\delta+d)}.$$

Proof of corollary 3.3.2. It suffices to note that the condition $\rho_n=O(a^{-n})$ implies that $\sum_{i=1}^{\infty} \rho(2^i)<\infty$, and then we can choose d_n to be constant and q_n to be proportional to logn. Therefore the term $q_n \log n(nh^d)^{-1}$ in W_n can be suppressed since it is of lower order than $(d_n \log^2 n M_n n^{-1} h^{-d})^{1/2}$. The balance between the two remaining terms in W_n is then realised by h^* defined in (3.3.15).

Corollary 3.3.3. Assume that the conditions of theorem 3.3.4 hold and that we have $\rho_n \leq l n^{-a}$ for some a>0 and l>0.

1st case: $a < (d+k+\delta)/(2k+2\delta)$. If the bandwidth is taken to be, for some finite positive constant C, of the form

$$h^* = C(n^{-2a/(2a+1)}\log n)^{1/(d+k+\delta)}, \tag{3.3.16}$$

then we get a rate of convergence in

$$W_n^* = (n^{-2a/(2a+1)}\log n)^{(k+\delta)/(2k+2\delta+d)}.$$

2nd case: $a \geq (d+k+\delta)/(2k+2\delta)$. If the bandwidth is taken to be, for some finite positive constant C, of the form

$$h^* = C\,(n^{-1}\log^2 n)^{1/(2k+2\delta+d)}, \tag{3.3.17}$$

then we obtain a rate of convergence in

$$W_n^* = (n^{-1}\log^2 n)^{(k+\delta)/(2k+2\delta+d)}.$$

Proof of corollary 3.3.3. As in corollary 3.3.1 we have $\sum_{i=1}^{\infty} \rho(2^i)<\infty$, and then we can choose d_n to be constant and $q_n=(4l^2 n)^{1/(2d+1)}$. Therefore the term $(d_n \log^2 n M_n n^{-1} h^{-d})^{1/2}$ (resp. the term $q_n \log n(nh^d)^{-1}$) in W_n can be suppressed in the first case (resp. the second case) and then the balance between the two remaining terms is realised by h^* defined in (3.3.16) (resp. in (3.3.17)).

Proof of theorem 3.3.4. The proof is performed over the same steps that for theorem 3.3.3, and it is only necessary to check a result similar to lemma 3.3.1 since the other computations in proof of theorem 3.3.3 did not concern the dependence structure and can therefore be applied to the ρ-mixing case investigated herein. So we only have to prove the following lemma.

Lemma 3.3.2. Under the conditions of theorem 3.3.2 we have for some $\epsilon>0$

$$\sum_{n=1}^{\infty} P(W_n \sup_{x\in G} |\bar{g}_n(x) - E\bar{g}_n(x)| > \epsilon) < \infty.$$

Proof of lemma 3.3.2. Let us denote along this proof by C a generic constant. Using the same decompositions as in the case of φ-mixing variables (formulas (3.3.9) and (3.3.10) in proof of theroem 3.3.3), it will be enough to show that

$$\sum_{n=1}^{\infty} P(h^{k+\gamma} \sup_{x\in G} |\hat{S}_n(x)| > \epsilon) < \infty, \tag{3.3.18}$$

and

$$\sum_{n=1}^{\infty} P(W_n \max_{k=1,..l_n} |S_n(t_k)| > \epsilon) < \infty. \tag{3.3.19}$$

The proof of (3.3.12), (which was the equivalent of (3.3.18) in φ-mixing case), did not use any condition on the dependence structure between the observations, and is therefore directly appliable here. So, we have only to prove (3.3.19).

The main tool of this proof is the application of an exponential inequality for ρ-mixing variables, for instance that given by Peligrad (1987, corollary 3.4) which is recalled in theorem 2.2.2. Let us denote by

$$\Psi_i = h^{-d/2} Y_i I_{\{|Y_i|\le M_n\}} K((x-X_i)/h),$$

and

$$Z_i = \Psi_i - E\Psi_i .$$

Using (3.3.7) and theorem 2.2.3, we are within the conditions of application of theorem 2.2.2, and using similar notations we have

$$d = (Cnd_n)^{1/2},\ D = (Cnd_n h^d M_n^{-2})^{-1/2}$$

and

$$\alpha = (5A+4q_n(Cnd_n h^p)^{-1/2})^{-1}\log(2\mu-1).$$

So, that gives for any sequence $(\epsilon_n)_N$

$$P(|\bar{g}_n(x)-E\bar{g}_n(x)|>\epsilon_n) = P(n^{-1}h^{-d/2}|\sum_{i=1}^{n} Z_i| > \epsilon_n)$$

$$= P(Cn^{-1/2}h^{-d/2}d_n^{1/2}\alpha|\sum_{i=1}^{n} Z_i|(nd_n)^{-1/2}\geq\alpha\epsilon_n)$$

$$\leq U(\mu)\exp(-C\epsilon_n W_n^{-1}\log n).$$

Fixing now $\epsilon>0$, and choosing $\epsilon_n = \epsilon W_n^{-1}$, we get

$$\sup_{x\in\mathbb{R}^d} P(W_n|\bar{g}_n(x) - E\bar{g}_n(x)|>\epsilon) \leq U(\mu)n^{-C\epsilon}.$$

So we have

$$P(W_n \max_{k=1,\dots,l_n} |S_n(t_k)| > \epsilon) \leq l_n \sup_{x\in G} P(W_n|S_n(x)| > \epsilon)$$

$$\leq l_n U(\mu)\, n^{-C\epsilon}.$$

It suffices to take ϵ big enough and to use the decomposition (3.3.9) to get (3.3.19), which completes the proof of this lemma, and therefore also completes the proof of theorem 3.3.4.

<u>Proofs of remarks 3.3.6 and 3.3.7.</u> The arguments used to prove remarks 3.3.4 and 3.3.5 also apply here.

3.5. Case of α-mixing variables.

Let us assume through the following section that the random variables $(X_i,Y_i)_N$ are α-mixing, following the definition 2.2.3. The properties of the estimator r_n will depend on the mixing coefficients $(\alpha_n)_N$ through an increasing sequence $(s_n)_N$ of integers such that

$$\exists A<\infty,\ \forall n\in N,\ 1 \le s_n \le n/2 \text{ and } n\alpha_{s_n}^{(2s_n/3n)}/s_n \le A. \qquad (3.3.20)$$

Let us recall that M_n is an increasing sequence of real numbers satisfying

$$M_n = n^{\xi},\ \text{for some } \xi \in (4(\beta+2)^{-1},\ 1). \qquad (3.3.21)$$

Theorem 3.3.5. Uniform convergence. Assume that (X_i, Y_i) is α-mixing and that the conditions (A.1)–(A.3) and (K.1)–(K.4) hold. If the function r is continuous on G and if the bandwidth h is such that

$$nh^d/(s_n M_n \log n) \longrightarrow \infty, \qquad (H.5)$$

then we have

$$\sup_{x\in G} |r_n(x) - r(x)| \xrightarrow{co.} 0.$$

Note that the proof of this theorem follows directly from theorem 3.3.6 below.

The following result specifies the rate of convergence of the estimate r_n as a function of the smoothness of the regression function r.

Theorem 3.3.6. Rate of convergence. Assume that (X_i, Y_i) is α-mixing and that conditions (A.1)–(A.4) and (K.1)–(K.6) hold. If the bandwidth is such that the sequence

$$S_n = h^{k+\gamma} + [(s_n M_n \log n)/(nh^d)]^{1/2},$$

satisfies the condition

$$S_n \longrightarrow 0 \text{ as } n \longrightarrow \infty, \qquad (H.6)$$

then we have

$$\sup_{x\in G} |r_n(x) - r(x)| = O(S_n),\ \text{co.}$$

Before giving the proof of this result let us give additional results.

Remark 3.3.8. If in place of (A.2) we assume

$$\exists M<\infty,\ \exists \alpha>0,\ \forall i\in N,\ E\exp(\alpha Y_i) \le M, \qquad (A.2.a)$$

we get a rate of convergence in

$$S_n = h^{k+\gamma} + [(s_n \log^2 n)/(nh^d)]^{1/2}.$$

Remark 3.3.9. If in place of (A.2) we assume

$$\exists M<\infty,\ \forall i\in N \quad |Y_i| \leq M, \qquad \text{(A.2.b)}$$

we get the rate of convergence

$$V_n = h^{k+\gamma} + [(s_n \log n)/(nh^d)]^{1/2}.$$

These rates of convergence can be optimised by a suitable asymptotic choice of the bandwidth h.

Corollary 3.3.4. Assume that the conditions of theorem 3.3.6 hold and that the bandwidth is taken to be, for some finite positive constant C, of the form

$$h^* = C\ (n^{-1} s_n M_n \log n)^{1/(2k+2\gamma+d)}, \qquad (3.3.22)$$

then we obtain a rate of convergence in

$$S_n^* = (n^{-1} s_n M_n \log n)^{(k+\gamma)/(2k+2\gamma+d)}.$$

Proof of remarks 3.3.8, 3.3.9 and corollary 3.3.4. These proofs are omitted since they are exactly the same as those of remarks 3.3.4, 3.3.5 and corollary 3.3.1.

Proof of theorem 3.3.6. The structure of the proof is similar to those of the proof of theorem 3.3.2 above. Let us use the same notations. For the same reasons, we only have to prove that

$$\sup_{x\in G} A_1(x) = O((n^{-1} h^{-d} s_n M_n \log n)^{1/2}), \qquad (3.3.23)$$

where

$$A_1 = g_n - Eg_n.$$

As before we decompose g_n in the following way

$$g_n^+(x) = (nh^d)^{-1} \sum_i Y_i K((x-X_i)/h)\ I_{\{|Y_i|\geq M_n\}},$$

and

$$g_n^-(x) = g_n(x) - g_n^+(x).$$

The result (3.3.6) that was stated along the proof of theorem did not use the dependence structure between the variables (X_i,Y_i). Therefore it is still valid in our case. Let us recall that this result stated that there exists some $\epsilon>0$ such that

$$P(M_n^{-1}nh^d \sup_{x\in\mathbb{R}^d} |g_n^+(x) - Eg_n^+(x)|\geq\epsilon) \leq LnM_n^{-1-\beta/2} < \infty.$$

Finally, we complete the proof of theorem 3.3.6 by using this inequality together with lemma 3.3.3 below.

<u>**Lemma 3.3.3.**</u> Under the conditions of theorem 3.3.6 we have for any $\epsilon>0$,

$$\sum_{n=1}^{\infty} P(S_n \sup_{x\in G} |g_n^-(x) - Eg_n^-(x)| > \epsilon) < \infty.$$

<u>**Proof of lemma 3.3.3.**</u> Let us denote along this proof by C a generic constant. The main tool of this proof is the application of the exponential inequality of Carbon (1983) which is stated in theorem 2.2.6. Let us use the same notations as those of theorem 2.2.6., and denote by

$$\Psi_i = n^{-1}h^{-d}Y_i I_{\{|Y_i|\leq M_n\}},$$

and

$$Z_i = \Psi_i - E\Psi_i.$$

Following the same arguments as for the proof of lemma 3.3.1 we have,

$$|Z_i| \leq n^{-1}h^{-d}M_n\bar{K},$$

and

$$EZ_i^2 \leq Cn^{-2}h^{-d}\int(r^2(u)+V)h^{-d}K((x-u)/h)du \leq C\hat{K}n^{-2}h^{-d}.$$

Take $k=s_n$ (resp. $k=s_0$) if $\lim s_n = \infty$ (resp. if $\exists n_0, \forall n>n_0\ s_n=s_0$), and choose α such that

$$\alpha = Cnh^d/(kM_n) \text{ and } (\alpha kn^{-1}h^{-d}M_n\bar{K}) > e/4.$$

Applying now theorem 2.2.6 we get for any sequence $(\epsilon_n)_N$

$$P(|g_n^-(x)-Eg_n^-(x)|>\epsilon_n) \leq C_1\exp(-C_2nh^d\epsilon_n^2/(s_nM_n)),$$

for some constant C_1 and C_2 independent of x. That leads to

$$\sup_{x\in\mathbb{R}^d} P(|\bar{g}_n(x)-E\bar{g}_n(x)|>\epsilon_n) \leq C_1\exp(-C_2nh^d\epsilon_n^2/(s_nM_n)).$$

By Lipschitz considerations, we go from this inequality to lemma 3.3.3 following exactly the same steps as those used to go from (3.3.8) to lemma 3.3.1. This completes the proof of lemma 3.3.3 and therefore also the proof of theorem 3.3.6.

4. Time series analysis and prediction under mixing conditions.

4.1. Introduction.

The introduction of a dependence stucture is of particular interest in the setting of regression estimation because the possible applications to the estimation of the autoregression function of a process, problems in which an independence assumption is clearly not acceptable. In a first approach we will show how the results obtained in section 3.3 on kernel regression estimates lead to similar results on nonparametric autoregression estimates and on kernel predictors. Then we will investigate, as an application of autoregression function estimation, the problem of the prediction of future values of a Markov process. Finally we will investigate in details the case of autoregressive processes.

Such an approach of nonparametric time series analysis and prediction problems has mainly been investigated by Collomb (1984) and (1985b), Sarda and Vieu (1985b and 1988b)), Collomb and Härdle (1986) and in some recent paper by Truong and Stone (1988).

4.2. Nonparametric Time Series Analysis.

Let $(Z_n)_N$ be a process valued in $\mathbb{R}^\delta$. An important tool to study the dependence between future and past values of the process is the estimation of the autoregression functions, defined for q and s positive integers by

$$R_{q,s}(u) = E(Z_{n+s}|(Z_{n-q+1},\ldots,Z_n)=u),\ \forall n\geq q.$$

In the following we assume that such a function exists, and we will abreviate it from now on in R when no confusion will be possible. We note that such a condition is satisfied as soon as the process $(Z_n)_N$ is stationary, but the stationarity of the process will not be necessary in a lot of the results presented here. The kernel autoregression estimates have been defined by Collomb (1984) from the Watson-Nadaraya kernel estimates (3.2.1). These estimates are of the form

$$R_n(u) = \frac{\sum_{i=q}^{n-s} Z_{i+s}K((u-(Z_{i-q+1},\ldots,Z_i))/h)}{\sum_{i=q}^{n-s} K((u-(Z_{i-q+1},\ldots,Z_i))/h)},\ \forall u\in\mathbb{R}^{\delta q}.$$

The first three theorems state, under appropriate asymptotic behaviour of the bandwidth h, the uniform convergence of R_n on some compact set. The other next three theorems will specify the rate of convergence as a function of the smoothness of the true function R. In the following G will be a compact subset of $\mathbb{R}^{\delta q}$, and the sequences (m_n), (d_n), (q_n), (s_n) and (M_n) are those defined respectively by (3.3.1), (3.3.13), (3.3.14), (3.3.20) and (3.2.2).

Theorem 3.4.1. Assume that the conditions (A.1)-(A.3) are satisfied for the random variables

$$X_i = (Z_i,\dots,Z_{i+q-1}) \text{ and } Y_i = Z_{i+q+s-1}, \tag{3.4.1}$$

and for $d=\delta q$. Assume that the kernel function satisfies (K.1)-(K.4), that R is continuous on G and that $(Z_n)_N$ is **φ-mixing**. If the bandwidth h satisfies

$$nh^{\delta q}/(m_n M_n \log n) \longrightarrow \infty, \tag{H.7}$$

then we have

$$\sup_{x\in G} |R_n(x) - R(x)| \xrightarrow{co.} 0. \tag{3.4.2}$$

Theorem 3.4.2. Assume that the conditions (A.1)-(A.3) are satisfied for the random variables defined in (3.4.1) and for $d=\delta q$. Assume that the kernel function satisfies (K.1)-(K.4), that R is continuous on G and that $(Z_n)_N$ is **ρ-mixing**. If the bandwidth h satisfies

$$nh^{\delta q}/(d_n M_n \log^2 n) \longrightarrow \infty \text{ and } nh^{\delta q}(q_n \log n) \longrightarrow \infty, \tag{H.8}$$

then we have

$$\sup_{x\in G} |R_n(x) - R(x)| \xrightarrow{co.} 0. \tag{3.4.3}$$

Theorem 3.4.3. Assume that the conditions (A.1)-(A.3) are satisfied for the random variables defined in (3.4.1) and for $d=\delta q$. Assume that the kernel function satisfies (K.1)-(K.4), that R is continuous on G and that $(Z_n)_N$ is **α-mixing**. If the bandwidth h satisfies

$$nh^{\delta q}/(s_n M_n \log n) \longrightarrow \infty, \tag{H.9}$$

then we have

$$\sup_{x\in G} |R_n(x) - R(x)| \xrightarrow{co.} 0. \tag{3.4.4}$$

Theorem 3.4.4. Assume that conditions of theorem 3.4.1 are satisfied and that in addition (A.4), (K.5) and (K.6) hold. Assume that the bandwidth h is such that

$$\hat{V}_n = h^{k+\gamma} + ((m_n M_n \log n/(nh^{\delta q}))^{1/2} \longrightarrow \infty, \qquad \text{(H.10)}$$

then we have

$$\sup_{x \in G} |R_n(x) - R(x)| = O(\hat{V}_n) \text{ co.} \qquad (3.4.5)$$

Theorem 3.4.5. Assume that conditions of theorem 3.4.2 are satisfied and that in addition (A.4), (K.5) and (K.6) hold. Assume that the bandwidth h is such that

$$\hat{W}_n = h^{k+\gamma} + (n^{-1} h^{-\delta q} d_n M_n \log^2 n)^{1/2} + n^{-1} h^{-\delta q} q_n \log n \text{-->} \infty, \qquad \text{(H.11)}$$

then we have

$$\sup_{x \in G} |R_n(x) - R(x)| = O(\hat{W}_n) \text{ co.} \qquad (3.4.6)$$

Theorem 3.4.6. Assume that conditions of theorem 3.4.3 are satisfied and that in addition (A.4), (K.5) and (K.6) hold. Assume that the bandwidth h is such that

$$\hat{S}_n = h^{k+\gamma} + ((s_n M_n \log n/(nh^{\delta q}))^{1/2} \longrightarrow \infty, \qquad \text{(H.12)}$$

then we have

$$\sup_{x \in G} |R_n(x) - R(x)| = O(\hat{S}_n) \text{ co.} \qquad (3.4.7)$$

Remark 3.4.1. These rates of convergence can be optimised by using a suitable bandwidth which balances the trade-off between the different components of the rates. The results are exactly those given in corollaries 3.3.1, 3.3.2, 3.3.3 and 3.3.4 with $d=\delta q$.

Remark 3.4.2. It is necessary to note that, even if theorems 3.4.2, 3.4.4 and 3.4.6 state more powerful results, theorems 3.4.1, 3.4.3 and 3.4.5 are applicable to a larger class of processes since they do not need to assume the stationarity of $(Z_n)_N$, while in theorems 3.4.2, 3.4.4 and 3.4.6 this assumption is automatically implied by (A.4) which assumes the existence of a common density. In particular, theorem 3.4.1 includes the autogressive processes,

$$Z_i = R(Z_{i-1},\dots,Z_{i-q}) + \epsilon_i,$$

for which a stationarity assumption is clearly not reasonable. This kind of processes will be treated in details below (see section 4.4 of this chapter).

<u>**Proof of theorems 3.4.1-3.4.3.**</u> These theorems are obvious consequences of theorems 3.4.4.-3.4.6.

<u>**Proof of theorems 3.4.4-3.4.6.**</u> These results are quite direct applications of convergence results stated in section 3.3, and it would be tedious to give all the deatils. Therefore, we just sketch the proofs. Even if it is only dealing with φ-mixing processes, the paper of Collomb (1984) gives all the details of the following computations.
The first thing is to note that when the process $(Z_n)_N$ satisfies some mixing condition, the variables $(X_i,Y_i)_N$ defined in (3.4.1) satisfy the same condition with the same mixing coefficients.

The main difficulty is that we cannot directly apply the results of section 3.3 since there the variables Y_i were assumed to be univariate, which is not the case here. Therefore, we first show results on the function $\hat{R}$ defined from some measurable real valued function g by

$$\hat{R}(u) = E(g(Z_{n+s})|(Z_{n-q+1},\dots,Z_n)=u).$$

The estimator $\hat{R}_n$ of R is naturaly defined by changing Z_{i+s} in $g(Z_{i+s})$ in the definition of R_n, i.e.,

$$\hat{R}_n(u) = \frac{\sum_{i=q}^{n-s} T_{i+s}K((u-(Z_{i-q+1},\dots,Z_i))/h)}{\sum_{i=q}^{n-s} K((u-(Z_{i-q+1},\dots,Z_i))/h)},\ \forall u\in\mathbb{R}^{\delta q},$$

where

$$T_i = g(Z_i),\ \text{for any } i.$$

Theorem 3.3.2 (resp. 3.3.4, resp. 3.3.6) gives

$$\sup_{x\in G} |\hat{R}_n(x) - \hat{R}(x)| = O(Q_n)\ \text{co.},$$

where

$$Q_n = \hat{V}_n\ (\text{resp. } Q_n = \hat{W}_n,\ \text{resp. } Q_n = \hat{S}_n)$$

under the conditions of theorem 3.4.4 (resp. 3.4.5, resp. 3.4.6). Now, it suffices then to apply this result δ-times, choosing each time g to be the coordinate function

$$g_j(u_1,\dots u_\delta) = u_j, \text{ for } j=1,\dots,\delta,$$

to complete the proofs.

4.3. Markovian prediction.

Assume in this section that $(Z_n)_N$ is a stationary Markov process of order q, and denote by $(\Omega,\mathcal{A})$ its probability states space. For $m\in N^*$, for $x\in\Omega$ and for $A\in\mathcal{A}$, denote by $P(x,A)$ the transition probability function and by $P_m(x,A)$ the m-step transition probability function of this process. Let denote by υ a probability measure invariante with respect to $P(.,.)$.

Definition 3.4.1. (Doob, 1953, p.192). The process $(Z_n)_N$ is said to satisfy Doeblin's condition if there exists a finite probability measure Ψ defined on $\mathcal{A}$, a positive integer m and some $\epsilon>0$ such that we have

$$\Psi(\Omega) > 0,$$

and for any $A\in\mathcal{A}$

$$\Psi(A) \leq \epsilon \Rightarrow \forall x\in\Omega,\ P_m(x,A) \leq 1-\epsilon. \qquad \text{(D)}$$

Definition 3.4.2. (Rosenblatt, 1971b, p.206). The process $(Z_n)_N$ is said to satisfy the L^p norm condition $(1\leq p\leq\infty)$ if we have

$$\sup_{f\ 1} \frac{\|T_n f\|_p}{\|f\|} \longrightarrow 0 \text{ as } n\to\infty,$$

T being the transition operator on $L^p(d\upsilon)$ induced by $P(.,.)$. The L^p norm condition is often called G_p condition (Rosenblatt, 1980, Yakowitz , 1985b). For stationary Markov processes the G_2 condition has the following equivalent form: assume an interger ≥ 2 and real $0 < \alpha < 1$ such that for all bounded functions h with $E(h(Z_1)) = 0$ we have

$$E(E(h(Z_n) \mid Z_1)^2) \leq \alpha\, E(h(Z_1)^2).$$

The relations between these definitions and the mixing conditions can be seen through the following results.

Lemma 3.4.1. (Rosenblatt, 1971b, p.209).
i). If $(Z_n)_N$ satisfies Doeblin's condition then there exist $\gamma\in(0,\infty)$ and $\rho\in(0,1)$ such that for all $x\in\Omega$ and for all $A\in\mathcal{A}$,

$$|P_n(x,A) - \upsilon(A)| \leq \gamma\rho^n.$$

ii). If there exists a sequence (Z_n) such that for almost all x (w.r.t. υ)

$$\sup_{A\in\mathcal{A}} |P_n(x,A) - \upsilon(A)| \leq \gamma\rho^n, \tag{3.4.8}$$

then, there exists a set of υ-measure 1 on which D is satisfied.

Lemma 3.4.2. (Rosenblatt, 1971b, p.211). Let p and q be elements of $(1,\infty)$.
i) The following propositions are equivalent.

- $(Z_n)_N$ satisfies L^p condition;
- $(Z_n)_N$ satisfies L^q condition.

ii) The following propositions are equivalent.

- $(Z_n)_N$ satisfies L^1 condition;
- $(Z_n)_N$ satisfies L^∞ condition;
- $(Z_n)_N$ satisfies (3.4.8).

iii). If $(Z_n)_N$ satisfies L^1 condition then $(Z_n)_N$ satisfies L^2 condition, but these two conditions are not equivalent in general.

Lemma 3.4.3. (Rosenblatt, 1971b, p.212, formula 18). The two following assertions are equivalent

- $(Z_n)_N$ satisfies (3.4.8);
- $(Z_n)_N$ is φ-mixing.

All these lemmas lead to the following result.

Theorem 3.4.7. Let $(Z_n)_N$ be a stationary Markov process satisfying either D, L^1 or L^∞ condition. Then $(Z_n)_N$ is φ-mixing and its mixing coefficients (φ_n) are geometrically decaying, i.e. there exists $s\in(0,\infty)$ and $t\in(0,1)$ such that

$$\varphi_n \leq st^n. \tag{3.4.9}$$

Remark 3.4.3. To complete this result let us recall that Davydov (1975) showed that a Markov process can not be φ-mixing without satisfying (3.4.9).

Let us now suppose that one is interested in predicting Z_{n+s} from the n first realisations of the process $Z_1,\dots,Z_n$. Then, if the process is Markovian of order q, the best theoretical predictor (with respect to a quadratic measure of errors) is provided by the random vector $R(Z_{n-q+1},\dots,Z_n)$. The kernel predictor is defined from the autoregression function estimate R_n (see section 3.4.2) by $R_n(Z_{n-q+1},\dots,Z_n)$. We have the following results.

Theorem 3.4.8. (Collomb, 1984). Let $(Z_n)_N$ be a stationary Markov process of order q valued in a compact set G of $\mathbb{R}^\delta$, and satisfying either D, L^1 or L^∞. Assume that (A.1)-(A.3) are satisfied by the variables (X_i,Y_i) defined as in (3.4.1) and for $d=\delta q$ and that the kernel function satisfies (K.1)-(K.4). Assume that the bandwidth h satisfies

$$nh^{\delta q}/\log^2 n \longrightarrow \infty. \tag{H.13}$$

Then we have

$$|R_n(Z_{n-q+1},\dots,Z_n)-R(Z_{n-q+1},\dots,Z_n)| \xrightarrow{co.} 0.$$

Theorem 3.4.9. Assume that conditions of theorem 3.4.8 hold, and that in addition conditions (A.4), (K.5) and (K.6) are satisfied. Assume also that the bandwidth h satisfies

$$h^{k+\gamma} + (nh^{\delta q})^{-1/2}\log n \longrightarrow \infty. \tag{H.14}$$

Then we have

$$|(R_n-R)(Z_{n-q+1},\dots,Z_n)| = O(h^{k+\gamma}+(nh^{\delta q})^{-1/2}\log n),\ co.$$

Proof of theorem 3.4.8. Because of theorem 3.4.7 the process $(Z_n)_N$ is φ-mixing, and its coefficients are of the form

$$\varphi_n = st^n, \text{ for some } 0<s<\infty \text{ and some } 0<t<1.$$

Then the sequence $(m_n)_N$ defined in (3.3.1) can be taken such that for some finite real positive constant C

$$m_n = C\log n, \text{ for any } n\in N. \tag{3.4.10}$$

On the other hand, the variables Z_i are assumed to be uniformly bounded and so the sequence $(M_n)_N$ defined in (3.3.2) can be taken such that (see remark 3.3.5).

$$M_n = 1, \text{ for any } n\in N. \tag{3.4.11}$$

Because of (3.4.10) and (3.4.11) condition (H.13) is the same as condition (H.7). Applying then theorem 3.4.1 noting that for any random variable T valued in G we have

$$|R_n(T) - R(T)| \leq \sup_{x \in G} |R_n(x) - R(x)|, \qquad (3.4.12)$$

we complete the proof of theorem (3.4.8).

Proof of theorem 3.4.9. As before (3.4.10) and (3.4.11) imply that condition (H.14) is the same as condition (H.10). Therefore the proof follows from (3.4.12) and theorem 3.4.4.

Theorem 3.4.10. (Rosenblatt, 1971, pp.199-200). Assume that the stationary Markov process $(Z_i)_N$ satisfies the L^2 norm condition (G_2 condition). Then $(Z_i)_N$ is strongly mixing with geometrically decreasing mixing coefficients.

4.4. Case of autoregressive processes.

Let us in this section investigate the particular case of autoregressive processes taking values in $\mathbb{R}^\delta$. Such processes are of the following form

$$Z_i = R(Z_{i-1}, \ldots, Z_{i-q}) + \epsilon_i,$$

where q is some positive integer, $(\epsilon_i)_N$ is a sequence of i.i.d. zero mean variables and R is a function defined from $\mathbb{R}^{\delta q}$ to $\mathbb{R}^\delta$. Such processes are of Markov type but are clearly not stationary in most cases. In a series of papers Doukhan and Ghindes (1980a, 1980b, 1980c and 1983) present results about such processes. Among these results is the following lemma.

Lemma 3.4.4. (Doukhan and Ghindes, 1980a). Assume that $(Z_n)_N$ is an autoregressive process such that

R is bounded and the probability law of ϵ_1 is absolutely continuous w.r.t. Lebesgue measure. (DG)

Then this process satisfies (3.4.8).

After noting that Rosenblatt's proof of lemma 3.4.3 above were still valid without assuming the stationarity of the process, Collomb got directly from lemmas 3.4.3 and 3.4.4 the following result.

Theorem 3.4.10. (Collomb, 1984). Let $(Z_n)_N$ be an autoregressive process satisfying the condition (DG). Then $(Z_n)_N$ is φ-mixing with geometrically decreasing mixing coefficients (i.e. coefficients of the form of (3.4.9)).

Now we can apply the results of previous sections to the estimation of the function R. We have the following theorem.

Theorem 3.4.11. (Collomb, 1984). Let $(Z_n)_N$ be an autoregressive process of order q valued in a compact set G of $\mathbb{R}^\delta$, and satisfying the condition

(DG). Assume that (A.1)-(A.3) are satisfied by the variables (X_i, Y_i) defined as in (3.4.1) and for $d=\delta q$ and that the kernel function satisfies (K.1)-(K.4). Assume that the bandwidth h satisfies

$$nh^{\delta q}/\log^2 n \text{ -----> } \infty. \qquad \text{(H.13)}$$

Then we have

$$|R_n(Z_{n-q+1},\ldots,Z_n)-R(Z_{n-q+1},\ldots,Z_n)| \xrightarrow{co.} 0.$$

Proof of theorem 3.4.11. This proof is similar to that of theorem 3.4.8 above. It suffices to use theorem 3.4.10 (in place of theorem 3.4.7) to get formula (3.4.10). Results (3.4.11) and (3.4.12) are still valid. Finally theorem 3.4.11 follows from (3.4.10), (3.4.11), (3.4.12) and theorem 3.4.1.

Remark 3.4.4. We cannot in this case specify the rates of convergence, since we cannot apply theorem 3.4.4. The reason for this is that theorem 3.4.4 needs that the assumption (A.4) hold. This is not realised when the process is not stationary (since (A.4) implies the existency of a common density function).

5. Time series analysis for ergodic processes.

Let us consider, as in section 2 of this chapter, the problem of estimating

$$R(u) = E(Z_{n+s}|(Z_{n-q+1},\ldots,Z_n)=u),\ \forall n\geq q,$$

q and s being fixed positive integers. We study now the estimate R_n of R (defined in section 4.2 of this chapter) under less restrictive dependence structures than the mixing conditions, in order to involve a larger class of processes. We give a recent result (Delecroix, 1987) that states the uniform consistency of R_n when the process $(Z_i)_N$ is only assumed to be stationary and ergodic. Let us denote by C_0 the set of real valued continuous functions defined on $\mathbb{R}^{\delta q}$, and let us use the notation h(n) for the bandwidth in place of h as previously.

Theorem 3.5.1. (Delecroix, 1987, theorem 2, p.35). Assume that $(Z_n)_N$ is stationary and ergodic and that the conditional densities $g_{n,q}$ of $(Z_n,\ldots,Z_{n-q+1})$ given $(Z_{n-q},Z_{n-q-1},\ldots)$ exist, belong to C_0. Then under (A.2) and (K.1)-(K.4) we have on any compact subset G of $\mathbb{R}^{\delta q}$

$$\sup_{x\in G} |R_n(x) - R(x)| \xrightarrow{a.s.} 0,$$

as soon as the bandwidth satisfies

$$1-h(n)^{\delta q}h(n-2n^{1/2})^{-\delta q} = O(n^{-1/2}),$$

and

$$\exists\epsilon>0,\ n^{sq\delta-1/2}h(n^2)^{-2\delta^2q^2-3\delta q} = O(n^{-\epsilon}).$$

Proof of theorem 3.5.1. We refer to Delecroix (1987) for this proof. However, let us mention that this author needed to get this result the additional condition that

$$\exists g\in C_0,\ \| n^{-1}\sum_{i=1}^{n} g_{i,q}-g \|_{C_o} \xrightarrow{a.s.} 0. \tag{3.5.1}$$

In fact this assumption is not necessary since it is automatically satisfied. To see that, it suffices to note that C_0 is a separable Banach space and that $(g_{n,q})_{n\in N}$ is an ergodic and stationary sequence for any q (since $(Z_i)_N$ is also ergodic and stationary). Finally, (3.5.1) follows by using Beck's theorem (see theorem 2.1.1 above) and by taking

$$g = Eg_{1,q}.$$

Let us note that the techniques used by Delecroix to get this result are quite different than those used under mixing conditions and described in this book. As he pointed out, it could be very hard to specify the rate of convergence of R_n. Nevertheless the above theorem is a quite interesting result since it is one of the first convergence results stated under such an unrestrictive dependence condition (see also the related paper of Bosq and Delecroix, 1985).

6. Estimation of the derivatives of a regression.

Let us consider the formalisation of section 3 of this chapter and consider, in place of estimating the regression function r, the problem of estimating the function $D^t(r)$ where D^t is the differential operator

$$D^t = \frac{d^{\alpha_1}(x_1)\ldots d^{\alpha_d}(x_d)}{dx_1\ldots dx_d},\ \text{with } \sum_1^d \alpha_i = t,$$

assuming that the regression r has k (k>t) derivatives. Estimates of the derivatives of r provide us informations about the shape of r, and are also important tools in the determination of the "optimal theoretical bandwidths" since it is wellknown that the constants in the expressions of such bandwidth involve (among other things) the derivatives of r. What we call here the optimal theoretical bandwidths is some value of h that minimises some theoretical error (for example h^* described in the corollaries of section 3.3 is the optimal theoretical bandwidth for the L_U error-criterion).

Let us recall that the sequences (m_n), (d_n), (q_n), (s_n) and (M_n) are respectively defined by (3.3.1), (3.3.13), (3.3.14), (3.3.20) and (3.2.2).

Theorem 3.6.1. **φ-mixing case.** (Sarda and Vieu, 1988a). Assume that (A.1)-(A.4) hold and that the kernel function K has derivatives of order t satisfying (K.1)-(K.6), and that $(X_i,Y_i)_N$ are φ-mixing. If the bandwidth h is such that the sequence $\bar{V}_n$

$$\bar{V}_n = h^{k+\delta-t} + h^{-t}(n^{-1}h^{-d}m_nM_n\log n)^{1/2}$$

satisfies

$$\bar{V}_n \longrightarrow 0,$$

then we have

$$\sup_{x\in G} |D^t(r_n)-D^t(r)| = O(\bar{V}_n) \text{ co.}$$

Theorem 3.6.2. **ρ-mixing case.** Assume that (A.1)-(A.4) hold and that the kernel function K has derivatives of order t satisfying (K.1)-(K.6), and that $(X_i,Y_i)_N$ are ρ-mixing. If the bandwidth h is such that the sequence $\bar{W}_n$

$$\bar{W}_n = h^{k+\delta-t} + h^{-t}(n^{-1}h^{-d}d_nM_n\log^2 n)^{1/2} + h^{-t}(n^{-1}h^{-d}r_n\log n)$$

satisfies

$$\bar{W}_n \longrightarrow 0,$$

then we have

$$\sup_{x\in G} |D^t(r_n)-D^t(r)| = O(\bar{W}_n) \text{ co.}$$

Theorem 3.6.3. **α-mixing case.** Assume that (A.1)-(A.4) hold and that the kernel function K has derivatives of order t satisfying (K.1)-(K.6), and that $(X_i,Y_i)_N$ are α-mixing. If the bandwidth h is such that the sequence $\bar{S}_n$

$$\bar{S}_n = h^{k+\delta-t} + h^{-t}(n^{-1}h^{-d}s_nM_n\log n)^{1/2}$$

satisfies

$$\bar{S}_n \longrightarrow 0,$$

then we have

$$\sup_{x \in G} |D^t(r_n) - D^t(r)| = O(\bar{S}_n) \text{ co.}$$

<u>**Proof of theorems 3.6.1-3.6.3.**</u> These results can be stated exactly as theorems 3.3.3, 3.3.5 and 3.3.7. We omit therefore the proofs. In Sarda and Vieu (1988a) are given the techniques that are used to get theorem 3.6.1 just by slightly modifying the proof of theorem 3.3.3. Exactly the same techniques are applied to get theorem 3.6.2 (resp. theorem 3.6.3) by slight modification of the proof of theorem 3.3.3 (resp. 3.3.5.).

<u>**Remark 3.6.1.**</u> As in corollaries of section 3.3, these rates can be optimised by a suitable asymptotic bandwidth choice. In particular for independent pairs (take $m_n=1$) and when the variables Y_i are uniformly bounded (take $M_n=1$), it follows from theorem 3.7.1 that for h proportional to $(n\log n)^{1/(2k+2\gamma+d)}$ we reach the rate $(n\log n)^{(k+\gamma-t)/(2k+2\gamma+d)}$, which has been shown to be the optimal global rate of convergence by Stone (1982).

<u>**Remark 3.6.2.**</u> The main problem here, is that the derivatives of the Watson-Nadaraya kernel estimates are not very easy to compute since they are defined from a ratio. The estimates introduced by Gasser and Müller (1979), (and recently investigated under dependence by Hart, 1988), would be more apropriate in such problems, but unfortunately cannot be easily adapted to treat the case of random design points.

Chapter IV

DENSITY ESTIMATION.

1. Histogram.

If $(X_i)_N$ denotes an $\mathbb{R}^d$ valued stationary sequence, where the distribution of X_i is absolutely continuous and f denotes the density, then for the sample $X_1, X_2, \ldots, X_n$

$$f_n(x) = f_n(x, X_1, X_2, \ldots, X_n), \quad x \in \mathbb{R}^d$$

denotes the density estimate. Here f_n is a measurable function of its arguments and f_n is itself a density, i.e. for each $x \in \mathbb{R}^d$ $f_n(x) \geq 0$ and $\int f_n(x) \lambda(dx) = 1$. (In the sequel $\int g$ will denote the integral of g over $\mathbb{R}^d$ with respect to the Lebesgue measure λ.)

We consider the a.s. L_1 convergence of the histogram, the kernel estimate and the recursive kernel estimate. All these estimates are of form

$$f_n(x) = \frac{1}{n} \sum_{i=1}^{n} K_{ni}(x, X_i), \qquad (4.1.0)$$

where K_{ni}, n=1,2,... are measurable on $\mathbb{R}^{2d}$ and $K_{ni}(\cdot, y)$ are densities for all $y \in \mathbb{R}^d$. Because of the triangle inequality we have

$$\int |f_n(x) - f(x)| \lambda(dx)$$
$$\leq \int |f_n(x) - Ef_n(x)| \lambda(dx) + \int |Ef_n(x) - f(x)| \lambda(dx),$$

where the first term of the right hand side is stochastic (called "variance") and the second one is called bias. If the estimate f_n is of form (4.1.0) then the bias depends only on f, it does not depend on the multidimensional distribution of $X_1,X_2,\dots,X_n$, therefore it is the same as that of the bias for independent samples. In the sequel we deal only with the variance.

There is a natural question to answer: why do we concentrate here on L_1? Why do we deliberately exclude all commonly used error criteria like the sup norm, the L_2 norm, etcetera? Devroye and Györfi (1985a) and Devroye (1987) contain several arguments why it makes sense to consider L_1 here.

1.1. The a.s. L_1 consistency.

Let $P_n=\{A_{n1},A_{n2},\dots\}$ be a sequence of partitions such that $0<\lambda(A_{ij})$ for i=1,2,... and j=1,2,... where λ stands for the Lebesgue measure. If μ_n denotes the empirical measure for the sample $\{X_1,X_2,\dots,X_n\}$, then the histogram is defined by

$$f_n(x) = \frac{\mu_n(A_{ni})}{\lambda(A_{ni})}, \text{ if } x\in A_{ni}. \qquad (4.1.1)$$

If we introduce the notation

$$K_n(x,y) = \sum_{i=1}^{\infty} \frac{I_{A_{ni}}(x)\, I_{A_{ni}}(y)}{\lambda(A_{ni})} \text{ for any } x,y \in \mathbb{R}^d, \qquad (4.1.2)$$

then the equivalent definition of the histogram is

$$f_n(x) = \frac{1}{n}\sum_{i=1}^{n} K_n(x,X_i). \qquad (4.1.3)$$

For an independent sample we have the following: assume that for each set A of positive and finite Lebesgue measure and for each $\epsilon>0$ there is n_0 such that for all $n\ge n_0$ there is $A_n\in\sigma(P_n)$, such that

$$\lambda(A\Delta A_n) < \epsilon, \qquad (4.1.4)$$

(Δ denotes the symmetric difference). Moreover, assume that for all M>0 and for all spheres S

$$\limsup_{n\to\infty} \lambda\Big(\bigcup_{j:\lambda(A_{nj}\cap S)\le \frac{M}{n}} A_{nj}\cap S\Big) = 0. \qquad (4.1.5)$$

Then there is a universal constant c such that for all $\epsilon>0$ there is n_0 for which

$$P(\int|f_n(x)-f(x)|dx>\epsilon) \le \exp(-c\epsilon^2 n) \text{ for any } n\ge n_0, \quad (4.1.6)$$

(see Devroye and Györfi, 1985a, theorem 3.2).
The bias tends to 0 because of (4.1.4) (Abou-Jaoude, 1976a). The variance tends to 0 under the condition (4.1.5), since it is easy to see that

$$\int|f_n-Ef_n| = \sum_{i=1}^{\infty} |\mu_n(A_{ni})-\mu(A_{ni})|. \quad (4.1.7)$$

Therefore, for the exponential error bound of $P(|\int f_n-Ef_n|>\epsilon)$, lemma 3.2.1. is very important.

Instead of (4.1.5) in some cases we use other conditions like: assume that for each sphere S

$$\lim_{n\to\infty} \frac{1}{n} \#\{i \ ; \ A_{ni}\cap S\neq\Phi\} = 0, \quad (4.1.9)$$

or for a $\delta>2$ and for all spheres S

$$\lim_{n\to\infty} \sup \lambda\Big(\bigcup_{j:\lambda(A_{nj}\cap S)<n^{-1/\delta}} (A_{nj}\cap S)\Big) = 0. \quad (4.1.10)$$

For the classical definition of the histogram, P_n is a cubic partition where A_{ni} is a rectangle of the form

$$\prod_{j=1}^{d} [a_j k_{ij} h_n, \ a_j(k_{ij}+1)h_n),$$

$a_1,a_2,\dots,a_d$ are real and $\{k_{ij}\}$ are integers. Then (4.1.4) is equivalent to

$$\lim_{n\to\infty} h_n = 0,$$

while (4.1.5) and (4.1.8) are equivalent to

$$\lim_{n\to\infty} nh_n^d = \infty,$$

and (4.1.10) is equivalent to

$$\lim_{n\to\infty} nh_n^{\delta d} = \infty.$$

Theorem 4.1.1. Any of the following three assumptions implies that

$$\lim_{n\to\infty} \int |f_n - Ef_n| = 0 \text{ a.s.}$$

i. (4.1.5) is met and $(X_i)_N$ is φ-mixing;

ii. (4.1.9) is met and $(X_i)_N$ satisfies condition D1;

iii. $(X_i)_N$ is α-mixing, for $\delta_1 > 0$ we have

$$\sum_{i=1}^{U} i^{\delta_1} \alpha_i < \infty, \tag{4.1.11}$$

and for $\delta > \frac{1+\delta_1}{\delta_1}$ (4.10) holds.

Let us state some preliminary lemmas before giving the proof of this result.

Lemma 4.1.1. Let $P_n = \{A_{n1}, \ldots, A_{nk}\}$ be a partition and let F_n denote the σ-algebra generated by $(X_i)_{i \leq n}$, then for arbitrary integer $m > 0$ and $\epsilon > 0$ we have

$$P\Big(\sum_{i=1}^{k} \Big|\frac{1}{n}(I_{\{X_j \in A_{ni}\}} - P(X_j \in A_{ni}/F_{j-m}))\Big| > 2\epsilon\Big)$$
$$\leq 2^{k+1} m \exp\Big(-\frac{n}{2}\Big(\frac{\epsilon}{m}\Big)^2\Big). \tag{4.1.12}$$

Proof of lemma 4.1.1. For a Borel set A introduce the notation

$$Z_j = I_{\{X_j \in A\}},$$

and apply theorem 2.3.2. Then we have

$$P\Big(\Big|\frac{1}{n} \sum_{j=1}^{n} (I_{\{X_j \in A\}} - P(X_j \in A/F_{j-m}))\Big| > \epsilon\Big) < 2m\, e^{-\frac{\epsilon^2}{2nm^2}}. \tag{4.1.13}$$

Using again the trick of the proof of lemma 3.2.2. we get

$$P\Big(\sum_{i=1}^{k} \Big|\frac{1}{n} \sum_{j=1}^{n} (I_{\{X_j \in A_{ni}\}} - P(X_j \in A_{ni}/F_{j-m}))\Big| > 2\epsilon\Big)$$
$$\leq 2^k \sup_A P\Big(\frac{1}{n} \sum_{j=1}^{n} (I_{\{X_j \in A\}} - P(X_j \in A/F_{j-m})) > \epsilon\Big). \tag{4.1.14}$$

The proof is then complete by (4.1.13) and (4.1.14).

Lemma 4.1.2. If for density estimates f_n of form (4.0)

$$\lim \int |f_n - Ef_n| = 0 \text{ a.s.}$$

for all f of bounded support, then it holds for all f.

Proof of lemma 4.1.2. Given $(X_i)_N$ and f let $S_{0,R}$ be a sphere centered at 0 with radius R and let

$$T: R^d \longrightarrow S_{0,R+1}$$

be a one-to-one measurable transformation such that

$$Tx=x \text{ if } x \in S_{0,R}$$

and

$$Tx \notin S_{0,R} \text{ if } x \notin S_{0,R}.$$

Put $X_i^*=TX_i$ and let f^* be the density of X_i^*, f_n^* the density estimate from the sample $X_1^*,\dots,X_n^*$, then we have

$$\begin{aligned}
\int |f_n-f_n^*| &\le \frac{1}{n}\sum_{i=1}^{n} \int |K_{ni}(x,X_i) - K_{ni}(x,X_i^*)|dx \\
&\le \frac{1}{n}\sum_{i=1}^{n} I_{\{X_i \ne X_i^*\}} \int |K_{ni}(x,X_i) - K_{ni}(x,X_i^*)|dx \le \\
&\le \frac{1}{n}\sum_{i=1}^{n} I_{\{X_i \notin S_{0,R}\}} 2 = 2\,\mu_n(S_{0,R}^c).
\end{aligned}$$

Therefore, applying the condition of the lemma for f_n^* we get

$$\begin{aligned}
\limsup_{n\to\infty} \int |f_n - Ef_n| &\le \limsup_{n\to\infty} [\int |f_n-f_n^*|+\int |Ef_n-Ef_n^*|+\int |f_n^*-Ef_n^*|] \\
&\le \limsup_{n\to\infty} [2\mu_n(S_{0,R}^c)+2\mu(S_{0,R}^c)] \\
&= 4\mu(S_{0,R}^c) \text{ a.s.}
\end{aligned}$$

Since R was arbitrary lemma 4.1.2 is proved.

Proof of Theorem 4.1.1.

Proof of i. (Györfi, 1987). Because of lemma 4.1.2., without loss of generality, we may assume that there is a sphere S such that $\mu(S^c)=0$. Introduce the notations

$$H_n = \{i;\ \lambda(A_{ni}\cap S)>M/n\}, \qquad (4.1.15)$$

and

$$A_n = \bigcup_{i\notin H_n} A_{ni}, \qquad (4.1.16)$$

then on the one hand because of (4.1.10)

$$\int|f_n - Ef_n| \le \sum_{i\in H_n} |\mu_n(A_{ni})-\mu(A_{ni})| + |\mu_n(A_n)-\mu(A_n)| + 2\mu(A_n), \quad (4.1.17)$$

and on the other hand (4.1.5) implies $\lambda(A_n\cap S)\to 0$, and because of the absolutely continuity of μ, $\mu(A_n\cap S)\to 0$. Choose n_0 such that for all $n>n_0$

$$\mu(A_n\cap S) < \frac{\epsilon}{4},$$

then we get

$$2\mu(A_n) < \frac{\epsilon}{2}. \qquad (4.1.18)$$

Obviously we have

$$|H_n| \le \frac{\lambda(S)}{M/n},$$

and therefore, by suitable choice of M, we have

$$(|H_n|+1)/n < \frac{\epsilon}{64m_0}. \qquad (4.1.19)$$

Using (4.1.17), (4.1.18) and (4.1.19), lemma 3.2.2. implies that for $n\ge n_0$

$$P\left(\int|f_n-Ef_n|>\epsilon\right) \le P\left(\sum_{i\in H_n} |\mu_n(A_{ni})-\mu(A_{ni})+|\mu_n(A_n)-\mu(A_n)|>\frac{\epsilon}{2}\right)$$
$$\le \exp\left(-n\frac{\epsilon}{64m_0}\right), \qquad (4.1.20)$$

and the application of Borel-Cantelli lemma completes the proof of i.

Proof of iii. (Györfi, 1987). Put

$$H_n=\{i;\ \lambda(A_{ni}\cap S) > (1/n)^{1/\delta_2}\}, \qquad (4.1.21)$$

and A_n, S, n_0 as before, then

$$|H_n|+1 \le \frac{\lambda(S)}{(1/n)^{1/\delta_2}},$$

therefore

$$\frac{(|H_n|+1)^{\delta_2}}{n}$$

is bounded. Then, by lemma 3.2.4. for all $n \ge n_0$

$$P(\int |f_n - Ef_n| > \epsilon) = O(n^{-1-\delta_3}), \qquad (4.1.22)$$

and by the Borel-Cantelli lemma we are ready.

Proof of ii. Denote m the integer in the condition D1 and put

$$\tilde{f}_n(x) = \frac{1}{n} \sum_{j=1}^{n} E(K_n(x,X_j)/F_{j-m}), \qquad (4.1.23)$$

then

$$\begin{aligned} \|\tilde{f}_n - Ef_n\|_{L_1} &= \int |\int K_n(x,u)(\frac{1}{n} \sum_{j=1}^{n} g_j(u) - f(u)du|dx \\ &\le \int (\int K_n(x,u)dx) |\frac{1}{n} \sum_{j=1}^{n} g_j(u) - f(u)|du \\ &= \int |\frac{1}{n} \sum_{j=1}^{n} g_j(u) - f(u)|du, \end{aligned} \qquad (4.1.24)$$

where g_j denotes the conditional density of X_j given F_{j-m}. Then $(g_j)_N$ is a stationary and ergodic sequence taking values in L_1 which is separable. Therefore, by theorem 2.1.1., we get

$$\lim_{n\to\infty} \|\frac{1}{n} \sum_{j=1}^{n} g_i - f\| = 0 \text{ a.s.},$$

and so (4.1.24) implies that $\|\tilde{f}_n - Ef_n\| \to 0$ a.s. We have to prove that $\int |\tilde{f}_n - f_n| \to 0$ a.s. If we introduce the notation

$$\tilde{\mu}_n(A) = \frac{1}{n} \sum_{j=1}^{n} P(X_j \in A/F_{j-m}), \qquad (4.1.25)$$

then we have

$$\int|f_n-\tilde{f}_n| = \sum_{i=1}^{\infty} |\mu_n(A_{ni})-\tilde{\mu}_n(A_{ni})|. \qquad (4.1.26)$$

Let S be as before and put

$$H_n = \{i;\ A_{ni}\cap S\neq\phi\},$$

then, by condition (4.1.9), we have

$$\lim_{n\to\infty} \frac{1}{n} |H_n| = 0, \qquad (4.1.27)$$

moreover (4.1.26) implies that

$$\int|f_n-\tilde{f}_n|= \sum_{i\in H_n} |\mu_n(A_{ni})-\tilde{\mu}_n(A_{ni})| \text{ a.s.}$$

(4.1.28)

Therefore, because of (4.1.27), (4.1.28) and lemma 4.1.1, there is n_0 such that

$$P(\int|f_n-\tilde{f}_n| > \epsilon) \leq \exp(-\frac{n}{8}(\frac{\epsilon}{m})^2) \text{ for } n>n_0, \qquad (4.1.29)$$

and the proof of theorem 4.1.1 is complete.

1.2. Is the histogram L_1 consistent for stationary and ergodic samples?

After theorem 4.1.1 there is an obvious question whether the density can be estimated from a stationary and ergodic sample, or the histogram does converge to the density f in L_1. We expect a positive answer since for each Borel set A $\mu_n(A)\to\mu(A)$ a.s. if the sample is stationary and ergodic. Since for the partition $P_n=\{A_{n1},A_{n2},...A_{nk}\}$ the variance of a histogram is

$$\sum_{i=1}^{k} |\mu(A_{ni})-\mu_n(A_{ni})|, \qquad (4.1.30)$$

we have the following question: under what condition on k does (4.1.30) converge to 0 ?

Paul C. Shields has shown an interesting and also somewhat shocking example. Put k=2, namely there is a sequence of Borel sets A_n n=1,2,..., and a stationary and ergodic sequence $(X_i)_N$ such that

$$P(|\mu_n(A_n)-\mu(A_n)| \geq 1/2) > 1/8 \text{ for any n.} \qquad (4.1.31)$$

Let α be a fixed irrational number and X_1 a random variable which is uniformly distributed on [0,1], and

$$X_{n+1} = (X_n+\alpha) \bmod 1 \text{ for } n=1,2,\ldots, \tag{4.1.32}$$

then $(X_i)_N$ is stationary and ergodic. We apply Rohlin's theorem (see Shields, 1973): let $(\Omega,\mathcal{A},Q)$ be a probability space with an ergodic transformation T, then for each $0<\epsilon<1$ and integer $N>0$ there is $F\in\mathcal{A}$ such that $F, TF, \ldots, T^{N-1}F$ are disjoint and

$$Q\left(\bigcup_{i=0}^{N-1} T^iF\right) > 1-\epsilon.$$

Apply Rohlin's theorem, when $\Omega=[0,1]$, $Q=\mu$, which is uniform on [0,1], $N=4n$, $\epsilon=1/2$ and

$$Tx=(x+\alpha) \bmod 1.$$

Then, there is an F for which

$$\mu\left(\bigcup_{i=0}^{4n-1} T^iF\right) > 1/2. \tag{4.1.33}$$

Put

$$A_n = \bigcup_{i=0}^{2n-1} T^iF \text{ and } B_n = \bigcup_{i=0}^{n-1} T^iF.$$

Since T^iF, $i=0,1,\ldots,4n-1$, are disjoint and T is measure preserving, we have

$$\mu(A_n) \le 1/2 \tag{4.1.34}$$

and

$$\mu(B_n) > 1/8. \tag{4.1.35}$$

The definitions of A_n and B_n imply that on the event $\{X_1\in B_n\}$ all of X_1, TX_1, $T^2X_1,\ldots,T^{n-1}X_1$ belongs to A_n, therefore on the event $\{X_1\in B_n\}$ we have

$$\mu_n(A_n)=1. \tag{4.1.36}$$

From (4.1.34), (4.1.35) and (4.1.36) we get

$$P(|\mu(A_n)-\mu_n(A_n)|\ge 1/2) \ge P(X_1\in B_n) > 1/8.$$

1.3. The rate of convergence.

In the sequel we study the rate of convergence of the expected L_1 error for the histogram. The bias term is the same as for independent case. Therefore the rate of convergence of the bias can be derived under the same smoothness conditions on the density f as in the independent case. The stochastic term may have the same rate of convergence if there is a tail condition on f. A working condition is that

$$\int\sqrt{f} < \infty$$

(Devroye and Györfi, 1985a, chapter V). The question is to find conditions under which the rate of convergence of variance like terms is the same as the rate for independent data.

Theorem 4.1.2. For cubic partitions assume that

$$\lim_{n\to\infty} h_n = 0, \text{ and } \lim_{n\to\infty} nh_n^d = \infty,$$

and that f has compact support.

i. If $(X_i)_N$ is independent then

$$\limsup_{n\to\infty} \sqrt{n}h_n^{d/2}\, E\int|f_n - Ef_n| < \int\sqrt{f}.$$

ii. If $(X_i)_N$ is ρ-mixing then

$$\limsup_{n\to\infty} \sqrt{n}h_n^{d/2}\, E\int|f_n - Ef_n| < \int\sqrt{f}\sqrt{1+\sum_{j=1}^{U}\rho_j}.$$

iii. If $(X_i)_N$ is α-mixing and for some $\delta>0$

$$\sum_{i=1}^{U} i^{\delta}\alpha_i < \infty,$$

then

$$\limsup_{n\to\infty} h_n^d\sqrt{n}\, E\int|f_n - Ef_n| < \infty.$$

iv. If $(X_i)_N$ satisfies condition D2 and for the notation

$$g(x,y) = \sup_n \Big[\, h_n^d \sum_{k=2}^{n} |f_k(x,y) - f(x)f(y)|\Big],$$

we have

$$\int\int g(x,y)\lambda(dx)\lambda(dy)<\infty,$$

then

$$\limsup_{n\to\infty} \sqrt{n}h_n^{d/2}\, E\int|f_n-Ef_n|< \int\sqrt{f} + \int\sqrt{g(x,x)}\lambda(dx).$$

Proof of theorem 4.1.2. For arbitrary $(X_i)_N$, we have

$$E\int|f_n-Ef_n| = \sum_{i=1}^{U} E|\mu_n(A_{ni})-\mu(A_{ni})|$$
$$< \sum_{i=1}^{U} \sqrt{E|\mu_n(A_{ni})-\mu(A_{ni})|^2}. \tag{4.1.37}$$

Proof of i. Let S be a sphere containing the support of f. Then (4.1.37) implies that

$$E\int|f_n-Ef_n| < \sum_{i=1}^{\infty} \sqrt{\mu(A_{ni})/n} = \frac{1}{\sqrt{n}h_n^{d/2}} \sum_{i=1}^{n} \sqrt{\frac{\mu(A_{ni})}{\lambda(A_{ni})}}\,\lambda(A_{ni}).$$

We have to show that for density f of compact support

$$\lim_{n\to\infty} \sum_{i=1}^{n} \sqrt{\frac{\mu(A_{ni})}{\lambda(A_{ni})}} = \int\sqrt{f}. \tag{4.1.38}$$

Let $\hat{f}$ be a continuous density with the support contained in S and

$$\int|f-\hat{f}| < \epsilon/\lambda(S),$$

then obviously $\lim_{n\to\infty} h_n = 0$ implies

$$\lim_{n\to\infty} \sum_{i=1}^{n} \sqrt{\frac{\int_{A_{ni}}\hat{f}}{\lambda(A_{ni})}}\,\lambda(A_{ni}) = \int\sqrt{\hat{f}}.$$

Moreover

$$\left|\int\sqrt{\hat{f}} - \int\sqrt{f}\right| \le \int\left|\sqrt{\hat{f}} - \sqrt{f}\right| \le \int\sqrt{|\hat{f} - f|} \le \sqrt{\lambda(S)}\sqrt{\int|\hat{f}-f|} < \sqrt{\epsilon}$$

and

$$\left|\sum_{i=1}^{n}\sqrt{\frac{\mu(A_{ni})}{\lambda(A_{ni})}}\lambda(A_{ni})-\sum_{i=1}^{n}\sqrt{\frac{\int_{A_{ni}}\hat{f}}{\lambda(A_{ni})}}\lambda(A_{ni})\right| \le \sqrt{\lambda(S)}\sqrt{\sum_i \left|\int_{A_{ni}}\hat{f}-\mu(A_{ni})\right|}$$

$$\le \sqrt{\lambda(S)}\sqrt{\int|\hat{f}-f|} < \sqrt{\epsilon}.$$

Thus we have

$$\limsup_{n\to\infty}\left|\sum_{i=1}^{n}\sqrt{\frac{\mu(A_{ni})}{\lambda(A_{ni})}}\lambda(A_{ni}) - \int\sqrt{f}\right| \le 2\sqrt{\epsilon}.$$

Proof of ii. For a Borel set A we have

$$E|\mu(A)-\mu_n(A)|^2 = \frac{1}{n^2}\sum_{j,m}\sum E[(I_{\{X_j\in A\}}-P(X_j\in A))(I_{\{X_m\in A\}}-P(X_m\in A))]$$

$$\frac{1}{n}\mu(A)(1-\mu(A))+\frac{2}{n^2}\sum_j E[(I_{\{X_j\in A\}}-P(X_j\in A))(I_{\{X_1\in A\}}-P(X_1\in A))](n-j+1),$$

which leads to

$$E|\mu(A)-\mu_n(A)|^2 \le$$

$$\text{"}\ \frac{1}{n}\mu(A) + \frac{2}{n}\sum_j |E[(I_{\{X_j\in A\}}-P(X_j\in A))(I_{\{X_1\in A\}}-P(X_1\in A))]|. \quad (4.1.39)$$

For ρ-mixing $(X_i)_N$ we have

$$E((I_{\{X_j\in A\}}-P(X_j\in A))(I_{\{X_1\in A\}}-P(X_1\in A)))\le\rho_j\mu(A)(1-\mu(A)), \quad (4.1.40)$$

and (4.1.37), (4.1.39) and (4.1.40) imply that

$$E\int|f_n-Ef_n| < \sum_{i=1}^{U}\frac{1}{\sqrt{n}}\sqrt{\mu(A_{ni})(1+2\sum_{j=2}^{n}\rho_j)}.$$

Therefore, using the technique applied in the proof of i, we get that

$$\limsup_{n\to\infty}\sqrt{n}h_n^{d/2}\int|f_n-Ef_n| \le \int\sqrt{f}\sqrt{1+2\sum_{j=1}^{\infty}\rho_j}.$$

Proof of iii. Apply theorem 2.2.4 for the random variables

$$Z_i = I_{\{X_i\in A\}}-\mu(A),$$

we get

$$E|n(\mu_n(A)-\mu(A))| \leq (E(|n(\mu_n(A)-\mu(A))|^{2+2\delta})^{\frac{1}{2+2\delta}} \leq (Cn^{1+\delta})^{\frac{1}{2+2\delta}}.$$

Therefore we have for some real positive constant C

$$\sum_{i=1}^{\infty} E|\mu_n(A_{ni})-\mu(A_{ni})| \leq \frac{1}{n}\frac{\lambda(S)}{h_n^d} C^{\frac{1}{2+2\delta}} n^{\frac{1+\delta}{2+2\delta}} = \frac{\lambda(S)\, C^{\frac{1}{2+2\delta}}}{h_n^d\sqrt{n}}.$$

<u>**Proof of iv.**</u> We have

$$\left|\sum_{j=2}^{n} E((I_{\{X_j\in A\}}-P(X_j\in A))(I_{\{X_1\in A\}}-P(X_1\in A)))\right| \text{ "}$$

$$\leq \sum_{j=2}^{n} \int_A\int_A |f_k(x,y)-f(x)f(y)|\lambda(dx)\lambda(dy) \text{ "}$$

$$\leq h_n^d \int_A\int_A g(x,y)\lambda(dx)\lambda(dy).$$

Therefore, because of (4.1.37), (4.1.38) and (4.1.39), we have to show that

$$\lim_{n\to\infty}\sum_i \sqrt{\frac{1}{\lambda(A_{ni})^2}\int_{A_{ni}}\int_{A_{ni}} g(x,y)\lambda(dx)\lambda(dy)}\ \lambda(A_{ni}) = \int\sqrt{g(x,x)}\lambda(dx),$$

which is just a slight modification of the proof of (4.1.38).

2. Kernel estimates.

2.1. The a.s. L_1 consistency.

Let $K \in L_1$ be a nonnegative function for which

$$\int_{\mathbb{R}^d} K(x)dx=1.$$

K is called kernel and the kernel estimate is defined by

$$f_n(x) = \frac{1}{n} \sum_{i=1}^{n} \frac{1}{h_n^d} K\left(\frac{x-X_i}{h_n}\right), \qquad (4.2.1)$$

(Rosenblatt, 1956 and Parzen, 1962), where h_n, $n \geq 1$, is a sequence of positive numbers such that

$$\lim_{n\to\infty} h_n = 0 \qquad (4.2.2)$$

and

$$\lim_{n\to\infty} nh_n^d = \infty. \qquad (4.2.3)$$

The kernel estimate is a popular density estimate since for sufficiently smooth f (for example for twice differentiable f) its asymptotic bias is less than that of the histogram (see Devroye and Györfi, 1985, chapter V).

For independent samples we have the following (Devroye, 1983 and Devroye and Györfi, 1985a): under the conditions (4.2.2) and (4.2.3), for all f and $\epsilon>0$ there is n_0 such that

$$P(\int |f_n-f| > \epsilon) \leq \exp(-c\epsilon^2 n) \text{ for } n \geq n_0,$$

where c depends only on the kernel. If, instead of $P(\int |f_n-f|>\epsilon)$, the probabilities $P(|f_n(x)-f(x)|>\epsilon)$ or $P(\sup_x |f_n(x)-f(x)|>\epsilon)$ are studied, then in the exponential bound the exponent depends on h_n, too (see e.g. Foeldes, 1974, Rejto and Revesz, 1973 and Revesz, 1972).

The majority of papers, dealing with dependent samples, deal with the conditions under which the rate of convergence of $E|f_n(x)-Ef_n(x)|^2$ is the same as that of for independent samples.

We say that the rate of convergence is the same as that of for independent samples if

$$\lim_{n\to\infty} nh_n^d \, E|f_n(x)-Ef_n(x)|^2 = f(x) \int K^2. \tag{A}$$

We say that the rate of convergence is essentially the same as that of for independent samples if there is a constant C such that

$$\lim_{n\to\infty} nh_n^d \, E|f_n(x)-Ef_n(x)|^2 \leq C\, f(x). \tag{B}$$

In theorem 2 of Masry (1983) it is shown that if d=1 and $(X_i)_N$ is φ-mixing with

$$\sum_{i=1}^{\infty} \varphi_i^{1/2} < \infty, \tag{4.2.4}$$

then the rate of convergence is (B). If, in addition, $(X_i)_N$ satisfies condition D2 then the rate is given by (A). For ρ-mixing $(X_i)_N$ and under the condition

$$\sum_{i=1}^{\infty} \rho_i < \infty$$

the rate is (B), and if, in addition, $(X_i)_N$ satisfies condition D2 then the rate is (A) (Masry, 1983, theorem 4). If $(X_i)_N$ is α-mixing with

$$\sum_{i=1}^{\infty} \alpha_i^q < \infty \text{ for } 0<q<1, \tag{4.2.5}$$

then there is a constant C such that

$$\limsup_n \, n\, h_n^{1+q} E|f_n(x)-f_n(x)|^2 \leq C\, f(x)^{1-q}.$$

If we have condition D2 and (4.2.5) for $0<q<\frac{1}{2}$ then the rate is (A) (Masry, 1983, theorem 3). Rosenblatt (1971) and (1985) has shown that we have rate (A) if condition D3 holds. Castellana and Leadbetter (1986) extended this result, when in the notation of condition D2 and for

$$\beta_n = \sup_{x,y} \sum_{i=1}^{n} |f_i(x,y)-f(x)f(y)|,$$

we have

$$h_n \beta_n \to 0.$$

If $(X_i)_N$ is a stationary and ergodic Markov process then, under Doeblin's condition (see details at section III.4.3), Roussas (1969) and, under the G_2 condition, Rosenblatt (1970) proved that the rate is (A) and showed the a.s. convergence of f_n, too.

If $(X_i)_N$ is φ-mixing with (4.2.4) then Földes (1974) gave exponential bound on the tail probability of $\sup_x |f_n(x)-f(x)|$. Also, under (4.2.4), Bosq (1973) proved the a.s. uniform convergence of f_n.

Bradley proved the asymptotic normality of $f_n(x)$ for kernel and recursive kernel rules if the sample is of condition D2 and is ρ-mixing with

$$\sum_{i=1}^{\infty} \rho(2^i) < \infty.$$

<u>**Theorem 4.2.1.**</u> Any of the following three assumptions implies

$$\lim \int |f_n - Ef_n| = 0 \text{ a.s.}$$

i. (4.2.3) holds and $(X_i)_N$ is φ-mixing;

ii. (4.2.3) holds and $(X_i)_N$ satisfies condition D1;

iii. $(X_i)_N$ is α-mixing and for some $\delta_1>0$

$$\sum_{i=1}^{\infty} i^{\delta_1} \alpha_i < \infty, \qquad (4.2.6)$$

and for $\delta_2 > 2(1+\delta_1)/\delta_1$

$$\lim_{n\to\infty} nh_n^{\delta_2 d} = \infty.$$

<u>**Proof of theorem 4.2.1.**</u>

<u>**Proof of i.**</u> Let us copy the proof of lemma 3.2 in Devroye and Györfi (1985a), with the only modification that instead of lemma 3.2.1, we use lemma 3.2.2. By lemma 4.1.2, without loss of generality, we assume that the support of f belongs to a sphere S. For given $\epsilon>0$ choose reals M, L, $a_1,\dots,a_N$ and rectangles $A_1,\dots,A_N$ such that for

$$K^*(x) = \sum_{i=1}^{N} a_i I_{A_i}(x),$$

we have

$$\int|K-K^*|<\epsilon \text{ and } |K^*|\leq M \text{ and } K^*(x)=0, \text{ if } x\notin[-L,L]^d.$$

Let f_n^* denote the kernel estimate if in (4.2.1) K is replaced by K^*, then

$$\int|f_n-Ef_n| \leq 2\epsilon + \int|f_n^*-Ef_n^*|$$
$$\leq 2\epsilon + M \sum_{i=1}^{N} \frac{1}{h_n^d} \int|\mu(x+h_nA_i)-\mu_n(x+h_nA_i)|dx.$$

Therefore, we have an exponential bound for $P(\int|f_n-Ef_n|>\epsilon)$ depending on K if we have an exponential bound on

$$P(\frac{1}{h_n^d} \int|\mu(x+h_nA) - \mu_n(x+h_nA)|dx > \epsilon),$$

(4.2.7)

for all rectangles $A\subset\mathbb{R}^d$. Let P be a partition of $\mathbb{R}^d$ consisting of rectangles of size h_n/N. Then A is of form

$$A = \prod_{i=1}^{d} [x_i, x_i+a_i). \qquad (4.2.8)$$

If $\min_{1\leq i\leq d} a_i > \frac{2}{N}$ then put

$$A^*= \prod_{i=1}^{d} [x_i+\frac{1}{N}, x_i+a_i- \frac{1}{N}). \qquad (4.2.9)$$

Introduce the sets C_x and C_x^* such that

$$C_x = x+h_nA - \bigcup_{\substack{B\in P\\ B\subset x+h_nA}} B \subseteq x + h_n(A-A^*) \overset{\Delta}{=} C_x^* , \qquad (4.2.10)$$

and note that

$$\frac{1}{h_n^d}\int|\mu(x+h_nA)-\mu_n(x+h_nA)|dx \leq \frac{1}{h_n^d}\int \sum_{\substack{B\in P\\ B\subset x+h_nA\\ B\cap S\neq\phi}} |\mu(B)-\mu_n(B)|dx$$
$$+ \frac{1}{h_n^d} \int \sum_{\substack{B\in P\\ B\subset x+h_nA\\ B\cap S=\phi}} |\mu(B) - \mu_n(B)|dx$$

$$+ \frac{1}{h_n^d} \int (\mu(C_x^*) + \mu_n(C_x^*))dx. \qquad (4.2.11)$$

Applying the fact that for a probability measure υ and set C

$$\int \upsilon(x+hC)dx = \lambda(hC), \qquad (4.2.12)$$

we have that the third term of the right hand side of (4.2.11) is less than

$$\frac{2}{h_n^d} \lambda(h_n(A-A^*)) = 2\lambda(A-A^*) < \epsilon, \quad (4.2.13)$$

by appropriate choice of N. The first term of r.h.s. of (4.2.11) is less than

$$\lambda(A) \sum_{\substack{B\in P \\ B\cap S\neq\emptyset}} |\mu(B)-\mu_n(B)|, \qquad (4.2.14)$$

for which we can apply lemma 3.2.2. since the number of sets with $B\cap S\neq\emptyset$ is less than $(2RN/h+2)^d=o(n)$ by (4.2.3). The second term of r.h.s. of (4.2.11) is 0.

Proof of iii. The proof is the same as that of i. The only difference is that for $P(\int|f_n-Ef_n|>\epsilon)$ we prove a bound $n^{-(1+\delta_3)}$ by applying lemma 3.2.4 instead of lemma 3.2.2.

Proof of ii. Let m be the integer in the definition of condition D1 and introduce the notation

$$\tilde{f}_n(x) = \frac{1}{n} \sum_{i=1}^{n} E(\frac{1}{h_n^d} K(\frac{x-X_i}{h_n})/F_{i-m}), \quad (4.2.15)$$

then

$$\|\tilde{f}_n-Ef_n\|_1 = \int| \int \frac{1}{h_d^n} K(\frac{x-u}{h_n})(\frac{1}{n} \sum_{j=1}^{n} g_j(u) - f(u))du|dx$$

$$(4.2.16)$$

$$\leq \int(\int \frac{1}{h_d^n} K(\frac{x-u}{h_n})dx|\frac{1}{n} \sum_{j=1}^{n} g_j(u) - f(u)|du = \|\frac{1}{n} \sum_{j=1}^{n} g_j-f\|_1,$$

where g_j denotes the conditional density of X_j given F_{j-m}. Then $g_1, g_2, \ldots$ is a stationary and ergodic sequence of L_1 valued random variables, therefore by theorem 2.1.1 we have

$$\lim_{n\to\infty} \|\frac{1}{n} \sum_{j=1}^{n} g_j - f\|_1 = 0 \text{ a.s.} \qquad (4.2.17)$$

Therefore (4.2.16) implies that $\lim_{n\to\infty} \|\tilde{f}_n - Ef_n\|_1=0$ a.s. It remains to prove that $\lim_{n\to\infty} \int\|f_n - \tilde{f}_n\|_1=0$ a.s. For the notation of (4.1.25) we show that for a rectangle A

$$\lim_{n\to\infty} \frac{1}{h_n^d} \int |\mu_n(x+h_nA)-\tilde{\mu}_n(x+h_nA)|dx = 0 \text{ a.s.}$$

(4.2.18)

As in the proof of i we have

$$\frac{1}{h_n^d} \int |\mu(x+h_nA) - \tilde{\mu}_n(x+h_nA)|dx$$
$$\leq \frac{1}{h_n^d} \int \sum_{B\subset x+h_nA} |\mu_n(B) - \tilde{\mu}_n(B)|dx + \frac{1}{h_n^d} \int(\mu(C_x^*) + \tilde{\mu}_n(C_x^*))dx$$
$$\leq \lambda(A) \sum_{B\in P} |\mu_n(B)-\tilde{\mu}_n(B)| + 2\lambda(A-A^*). \qquad (4.2.19)$$

The second term of r.h.s. (4.2.19) can be small as in the proof of i. Let S be as before and put

$$P_n' = \{B;\ B\in P,\ B\cap S\neq\Phi\}. \qquad (4.2.20)$$

Thus

$$\sum_{B\in P} |\mu_n(B)-\tilde{\mu}_n(B)| = \sum_{B\in P_n'} |\mu_n(B)-\tilde{\mu}_n(B)| \text{ a.s.,}$$

(4.2.21)

and by (4.2.3) $|P_n'|/n\to 0$; therefore, by lemma 4.1.1, there is n_0 such that for any $n\geq n_0$

$$P(\sum_{B\in P} |\mu_n(B)-\tilde{\mu}_n(B)|>\epsilon) \leq \exp(-\frac{n}{8}(\frac{\epsilon}{m})^2),$$

(4.2.22)

and by Borel-Cantelli's lemma, the proof of ii and of theorem 4.2.1 is ready.

2.2. The rate of convergence.

Applying the technique of section 1.3 of the present chapter, we may get the rate of convergence of expected L_1 error of kernel estimates.

Theorem 4.2.2. Assume that f and K have compact support, K is bounded, and

$$\lim_{n\to\infty} h_n = 0 \text{ and } \lim_{n\to\infty} nh_n^d = \infty.$$

i. If $(X_i)_N$ is independent then

$$\limsup_{n\to\infty} \sqrt{n}h_n^{d/2}\, E\int|f_n-Ef_n| < \int\sqrt{f}\sqrt{\int K^2}\ .$$

ii. If $(X_i)_N$ is ρ-mixing then

$$\limsup_{n\to\infty} \sqrt{n}h_n^{d/2}\, E\int|f_n-Ef_n| < \int\sqrt{f}\sqrt{\int K^2}\sqrt{1+\sum_{j=1}^{U}\rho_j}.$$

iii. If $(X_i)_N$ is α-mixing and for a $\delta>0$

$$\sum_{i=1}^{\infty} i^{\delta}\alpha_i<\infty$$

then

$$\limsup_{n\to\infty} h_n^d\sqrt{n}\, E\int|f_n-Ef_n|<\infty.$$

iv. If $(X_i)_N$ satisfies condition D2 and if we have

$$\int\int g(x,y)\lambda(dx)\lambda(dy)<\infty,$$

where

$$g(x,y)=\sup_n\, (h_n^d \sum_{k=2}^{n} |f_k(x,y)-f(x)f(y)|),$$

then we have

$$\limsup_{n\to\infty} \sqrt{n}h_n^{d/2}\, E\int|f_n-Ef_n| < \int\sqrt{f}\sqrt{\int K^2} + \int\sqrt{g(x,x)}\lambda(dx).$$

Proof of theorem 4.2.2. The proof of i ii and iv is very similar to that of theorem 4.1.2. It is based on the inequalities

$$E\int|f_n-Ef_n| \le \int\sqrt{E|f_n-Ef_n|}$$

$$\leq \int \frac{1}{h_n^d} \left[\int \frac{1}{n} K^2\left(\frac{x-y}{h_n}\right)dy + \sum_{j=2}^{n} \left| EK\left(\frac{x-X_1}{h_n}\right)K\left(\frac{x-X_j}{h_n}\right) - EK\left(\frac{x-X_1}{h_n}\right)EK\left(\frac{x-X_j}{h_n}\right) \right| \right]^{\frac{1}{2}} dx.$$

In case of iii the theorem 2.2.4 is used again for the bounded random variables

$$Z_i = K\left(\frac{x-X_i}{h_n^d}\right) - EK\left(\frac{x-X_i}{h_n^d}\right).$$

3. Recursive kernel estimate.

Let K be a kernel and h_n, n=1,2,... be a sequence of positive numbers, then the recursive kernel estimate is defined by

$$f_n(x) = \frac{1}{n} \sum_{i=1}^{n} \frac{1}{h_i^d} K\left(\frac{x-X_i}{h_i}\right), \qquad (4.3.1)$$

(Wolverton and Wagner, 1969 and Yamoto, 1971).

For independent samples

$$\lim_{n\to\infty} h_n = 0 \qquad (4.3.2)$$

and

$$\lim_{n\to\infty} \frac{nh_n^d}{\log\log n} = \infty, \qquad (4.3.3)$$

imply

$$\lim_{n\to\infty} \int |f-f_n| = 0 \text{ a.s.}, \qquad (4.3.4)$$

(Devroye and Györfi, 1985, theorem 7.2).

If d=1 and $(X_i)_N$ is dependent then Masry (1986) proved that

$$nh_n E|f_n(x)-Ef_n(x)|^2 \qquad (4.3.5)$$

has the same limit as for independent samples under the following conditions:
if $(X_i)_N$ is α-mixing and satisfies condition D2 with

$$\sum_{k=1}^{\infty} \alpha_k^a < \infty \qquad (4.3.6)$$

for some $0<a<1/2$. (4.3.6) can be weakened if there is a sequence of integers $(c_n)_N$ for which

$$c_n \to \infty, \ c_n h_n \to 0 \ \text{ and } \ (1/h)^{1-\gamma} \sum_{k=c_n}^{\infty} \alpha_k^{1-\gamma} \to 0$$

for some $0<\gamma<1$. For a ρ-mixing sample, condition D2 is assumed and

$$\sum_{k=1}^{\infty} \rho_k < \infty. \qquad (4.3.7)$$

If $(X_i)_N$ is ρ-mixing and there is a positive, monotonically nonincreasing sequence $(u_n)_N$ for which

$$\sum_{n=1}^{\infty} (\log n)(\log\log n)\, u_n^2 \rho_n^2 < \infty,$$

then for $\delta>0$

$$\left[\frac{n h_n^d u_n^2}{(\log n)(\log\log n)^{1+\delta}}\right]^{1/2} (f_n(x)-Ef(x)) \longrightarrow 0 \text{ a.s.},$$

(Masry and Györfi, 1987).

If $(X_i)_N$ is α-mixing and for some $r>2$

$$\sum_{n=1}^{U} ((\log n)(\log n))^{1+\delta} \alpha_n^{1-\frac{2}{r}} \sum_{j=n}^{\infty} \frac{1}{j^2 h_j^{2d(1-\frac{1}{r})}} < \infty,$$

then $\lim_{n\to\infty} (f_n(x)-Ef_n(x)) = 0$ a.s. (Masry, 1987).

If (X_i) is stationary and ergodic and the conditional density g_n of X_n given $X_1,\ldots,X_{n-1}$ exists with

$$E\|g_n\|_2 < \infty,$$

then

$$\lim_{n\to\infty} \|f_n - f\|_2 = 0 \text{ a.s.},$$

(Györfi, 1981).

Theorem 4.3.1. (Györfi and Masry, 1988). If $K \in L_2$ and $\lim_{n\to\infty} h_n = 0$, then any of the following three conditions implies that

$$\lim_{n\to\infty} \int |f_n - Ef_n| = 0 \text{ a.s.}$$

i. $(X_i)_N$ is ρ-mixing and

$$\sum_{j=1}^{\infty} \frac{1}{j^2 h_j^d} < \infty, \tag{4.3.8}$$

and for some $\delta>0$

$$\sum_{n=1}^{\infty} (\log n)(\log\log n)^{1+\delta} \rho_n^2 \sum_{j=n}^{\infty} \frac{1}{j^2 h_j^d} < \infty; \tag{4.3.9}$$

ii. $(X_i)_N$ satisfies condition D1, (4.3.8) holds and

$$\sup_n \frac{1}{n} \sum_{i=1}^{n-1} i \int \left| \left(\frac{h_{i+1}}{h_i}\right)^d K\left(\frac{h_{i+1}}{h_i} x\right) - K(x) \right| dx < \infty; \tag{4.3.10}$$

iii. $(X_i)_N$ is α-mixing, (4.3.8) holds and for some $\delta>0$

$$\sum_{n=1}^{\infty} (\log n)(\log\log n)^{1+\delta} \alpha_n \sum_{j=n}^{\infty} \frac{1}{j^2 h_j^d} < \infty. \tag{4.3.11}$$

The condition (4.3.10) is met, for example, when K is radially nonincreasing, i.e. for each fixed x, $K(\lambda x)$ is monoton nonincreasing for $0<\lambda<\infty$ and

$$\sup_n \frac{1}{n} \sum_{i=1}^{n-1} i\left(1-\min\left(\frac{h_{i+1}}{h_i}, \frac{h_i}{h_{i+1}}\right)^d\right) < \infty \tag{4.3.12}$$

since for such K and $0<a<1$

$$\int |aK(ax)-K(x)| \le 2(1-a^d),$$

(Devroye and Györfi, 1985, p.186). Condition (4.3.12) holds for the usual choice of h_n: $h_n = Cn^{-b}$, where $0<b<1/d$ and $0\le C<\infty$.

<u>**Lemma 4.3.1.**</u> If f'_n and $\tilde{f}_n$ are density estimates such that

$$\|f'_n-\tilde{f}_n\|_2 \to 0 \text{ a.s.} \tag{4.3.13}$$

and

$$\|\tilde{f}_n-f\|_2 \to 0 \text{ a.s.} \tag{4.3.14}$$

then we have

$$\|f'_n-f\|_1 \to 0 \text{ a.s.} \tag{4.3.15}$$

<u>**Proof of lemma 4.3.1.**</u> Let $(T_i)_N$ be a partition of $\mathbb{R}^d$ such that $\lambda(T_i)=1$, $i=1,2,\ldots$, and let f'_n and f be densities, therefore

$$\|f'_n-f\|_1 = 2\int_{f\geq f'_n}(f-f'_n) = 2\sum_{i=1}^{\infty}\int_{\substack{T_i\\ f\geq f'_n}}(f-f'_n)$$

and

$$\int_{\substack{T_i\\ f\geq f'_n}}(f-f'_n) \leq \int_{T_i}|f-f'_n| \leq \int_{T_i}|f-\tilde{f}_n| + \sqrt{\int_{T_i}|\tilde{f}_n-f'_n|^2}$$

$$\leq \|f-\tilde{f}_n\|_1 + \|\tilde{f}_n-f'_n\|_2 \xrightarrow{\text{a.s.}} 0 \text{ for all } i,$$

and the convergence is dominated since

$$\int_{\substack{T_i\\ f\geq f'_n}}(f-f'_n) \leq \int_{T_i} f \text{ and } \sum_{i=1}^{\infty}\int_{T_i} f=1.$$

Therefore by the dominated convergence theorem, this lemma is proven.

<u>**Proof of Theorem 4.3.1.**</u>

<u>**Proof of i.**</u> Applying Corollary 2.3.2 for $H=L_2$ and for

$$Z_n = \frac{1}{h_n^d}K\left(\frac{\cdot-X_n}{h_n}\right) - \frac{1}{h_n^d}EK\left(\frac{\cdot-X_n}{h_n}\right), \tag{4.3.16}$$

we get

$$E\|Z_n\|_2^2 \leq E\|\frac{1}{h_n^d}K(\frac{\cdot-X_n}{h_n})\|_2^2 = \frac{1}{h_n^d}\int K^2.$$

(4.3.17)

Therefore by (4.3.8) and (4.3.9) we have

$$\|f_n-Ef_n\|_2 \xrightarrow{a.s.} 0. \qquad (4.3.18)$$

Because of $\lim_{n\to\infty} h_n=0$ we have

$$\|Ef_n-f\|_1 \longrightarrow 0, \qquad (4.3.19)$$

(Devroye and Györfi, 1985a, theorem 2.1). So, (4.3.18), (4.3.19) and lemma 4.3.1. imply that

$$\|f_n-f\|_1 \xrightarrow{a.s.} 0. \qquad (4.3.20)$$

Therefore

$$\|f_n-Ef_n\|_1 \xrightarrow{a.s.} 0. \qquad (4.3.21)$$

<u>**Proof of ii.**</u> Let m be the integer in the condition D1 and define

$$F_i = \begin{cases} F_0^i \text{ if } i>0 \\ \{0,\Omega\} \text{ otherwise,} \end{cases}$$

and

$$\tilde{f}_n(x) = \frac{1}{n}\sum_{i=1}^{n} E(\frac{1}{h_i^d}K(\frac{x-X_i}{h_i})/F_{i-m}). \qquad (4.3.22)$$

If g_i denotes the conditional density of X_i given F_{i-m}, then

$$\tilde{f}_n - Ef_n = \frac{1}{n}\sum_{i=1}^{n} C_i(g_i-f) \text{ a.s.,} \qquad (4.3.23)$$

where C_i: $L_1 \to L_1$ is a bounded linear operator defined by

$$C_i g(x) = \int \frac{1}{h_i^d} K(\frac{x-y}{h_i})g(y)dy \qquad (4.3.24)$$

Since $(X_i)_N$ is stationary and ergodic, $(g_i)_N$ is stationary and ergodic, too, therefore by theorem 2.1.1.

$$\|\frac{1}{n}\sum_{i=1}^{n}(g_i-f)\|_1 \xrightarrow{a.s.} 0. \qquad (4.3.25)$$

Thus the a.s. convergence of (4.3.23) follows from (4.3.25) and from the Toeplitz theorem in Banach space (Fritz, 1974) if

$$\sup_n (\|C_n\|_1 + \frac{1}{n}\sum_{i=1}^{n-1} i\|C_{i+1} - C_i\|_1) < \infty. \qquad (4.3.26)$$

For all $g \in L_1$ we get

$$\|C_n g\|_1 = \int|\int\frac{1}{h_n^d}K(\frac{x-y}{h_n})g(y)dy|dx \leq \int(\int\frac{1}{h_n^d}K(\frac{x-y}{h_n})dx)|g(y)|dy$$

and therefore we have

$$\|C_n g\|_1 = \int|g(y)|dy = \|g\|_1 \qquad (4.3.27)$$

That implies in particular

$$\|C_n\|_1 \leq 1. \qquad (4.3.28)$$

By the same argument we have

$$\|C_{i+1} - C_i\|_1 \leq \int|\frac{1}{h_{i+1}^d}K(\frac{x}{h_{i+1}}) - \frac{1}{h_i^d}K(\frac{x}{h_i})|dx$$
$$= \int|(\frac{h_{i+1}}{h_i})^d K(x\frac{h_{i+1}}{h_i}) - K(x)|dx, \qquad (4.3.29)$$

therefore by (4.3.28), (4.3.29) and (4.3.10) we have (4.3.26), and therefore

$$\|\tilde{f}_n - Ef_n\|_1 \xrightarrow{a.s.} 0. \qquad (4.3.30)$$

It remains to prove that

$$\|f_n - \tilde{f}_n\|_2 \xrightarrow{a.s.} 0. \qquad (4.3.31)$$

Apply Corollary 2.3.1 for

$$Z_n = \frac{1}{h_n^d}K(\frac{.-X_n}{h_n}) - \frac{1}{h_n^d}E(K(\frac{.-X_n}{h_n})/F_{n-m}),$$

(4.3.32)

and

$$c_n^2 = \left\|\frac{1}{h_n^d}K\left(\frac{\cdot}{h_n}\right)\right\|_2^2 \le \frac{1}{h_n^d}\int K^2, \qquad (4.3.33)$$

when $\psi_n = 0$ for $n>m$. Then, by (4.3.30), (4.3.31), (4.3.32) and lemma 4.3.1, we get

$$\|f_n - \tilde{f}_n\|_1 \xrightarrow{a.s.} 0. \qquad (4.3.34)$$

Proof of iii. Apply corollary 2.3.3 for the notation (4.3.16) and

$$c_n^2 = 4\|K\|^2/h_n^d.$$

We have to verify (2.3.18). For this note that:

$$\|Z_n\|^2 \le \frac{c_n^2}{2h_n^d\|K\|^2}\left(\left\|K\left(\frac{\cdot-X_n}{h_n}\right)\right\|^2+\left\|EK\left(\frac{\cdot-X_n}{h_n}\right)\right\|^2\right)$$

$$\le \frac{c_n^2}{2\|K\|^2}\left(\|K(\cdot-X_n)\|^2+\|EK(\cdot-X_n)\|^2\right) = c_n^2.$$

Chapter V

DISTRIBUTION AND HAZARD FUNCTIONS ESTIMATION.

1. Introduction.

Let X be a random variable valued in $\mathbb{R}^d$, $d \geq 1$, with distribution F and density f. We investigate, from a sample $X_1,\ldots,X_n$ drawn from f, the estimation of the hazard function, also called failure rate, defined for any x in $\mathbb{R}^d$, $F(x)<1$, as

$$g(x) = f(x)/(1-F(x)). \tag{5.1.1}$$

The reason why estimation of the hazard function is interesting can be seen in the following example. Suppose that the variable X is real valued (d=1), positive and f is continuous. Typically X is referred as the "lifetime of a component" and the aim is to evaluate the probability that this component breaks down between times t and $t+dt$ $(dt \geq 0)$, knowing that up to time t there is no breakdown. This quantity is given by

$$P(t<X<t+dt\,|X>t) = \frac{P(t<X<t+dt)}{P(X>t)}$$

$$= \frac{\int_t^{t+dt} f(u)du}{1-F(t)},$$

and by continuity of the density f we have

$$\lim_{dt\to 0} P[t<X<t+dt\,|X>t]/dt = \frac{f(t)}{1-F(t)} = g(t).$$

Thus, the function g indicates the "instantaneous probability" that a breakdown occurs at time t. In medicine it is used as the death or failure rate. Rice and Rosenblatt (1976) deal with a problem of geophysics (microearthquakes) in which $g(t)$ is the risk to have a new microearthquake at time t, 0 being the reference time of the last one.

Although the case of real valued X is the most important in practice, we will investigate the general multidimensional problem.

Several nonparametric estimates have been proposed in the literature concerning failure rate estimation (see Singpurwalla and Wong (1983) and

Hassani, Sarda and Vieu (1986) for surveys). A straightforward way to estimate the function g is to take a ratio of an estimate of f over an estimate of 1-F

$$g_n(.) = \frac{f_n(.)}{1-F_n(.)} . \tag{5.1.2}$$

Most of the results in the literature concern pointwise convergence in L_2 and L_∞ norms of estimates when the variables $(X_i)_{i=1,n}$ are independent. We investigate here this kind of estimate in the mixing setting and in the L_∞ point of view.

The estimates we will use for F and f are the empirical distribution function F_n and the kernel density estimate f_n (or the k_n-NN density estimate $\hat{f}_n$) respectively.

We will derive in section 4 (respectively 5) uniform convergence results for the hazard function estimate based on f_n (respectively $\hat{f}_n$) on a compact subset G of the support of 1-F in $\mathbb{R}^d$. Assuming that f is continuous on $\mathbb{R}^d$ and bounded away from zero on $\hat{G}$, where $\hat{G}$ is a compact ϵ-neighborhood of G in $\mathbb{R}^d$, we have the following inequalities

$$\begin{aligned} & C_1 \sup|f_n(x)-f(x)|+C_2 \sup|F_n(x)-F(x)| \leq \\ & \leq \sup |g_n(x)-g(x)| \leq \qquad (5.1.3) \\ & \leq C_3 \sup |f_n(x)-f(x)| + C_4 \sup |F_n(x)-F(x)|, \end{aligned}$$

where C_1, C_2, C_3 and C_4 are positive constants and the supremum is taken over G. The same inequalities hold for the estimate $\hat{g}_n$ based on $\hat{f}_n$.

Thus, the results derived in hazard function estimation can be seen as applications of related results in distribution and density estimations. Therefore, in section 2 we will first give extensions of Glivenko-Cantelli's theorem for mixing processes, and in section 3 we will give uniform convergence results for the kernel density estimate defined by (4.2.1) above.

We note that the hazard function can be estimated directly using order statistics (see the surveys mentioned above). Nevertheless, we do not deal here with this kind of estimate.

Another point which is not treated here in this chapter is the selection of the smoothing parameter in practical situations. This problem will be studied in great detail in chapter VI.

2. Glivenko-Cantelli's theorem for mixing sequences.

The well-known empirical distribution function F_n is defined for any x in $\mathbb{R}^d$ as

$$F_n(x) = n^{-1} \sum_{i=1}^{n} I_{[X_i \leq x]}. \tag{5.2.1}$$

In this definition the order on $\mathbb{R}^d$ is defined for $u=(u_j)_{j=1}^d$ and $v=(v_j)_{j=1}^d$ by

$$u \leq v \iff u_j \leq v_j, \; j=1,\dots,d.$$

Let us recall that G is a compact subset of the support of 1-F in $\mathbb{R}^d$ and $\hat{G}$ a compact ϵ-neighborhood of G in $\mathbb{R}^d$. Then it exists a real $\epsilon>0$ such that for any x in G we have

$$F(x) \leq 1-\epsilon. \tag{5.2.2}$$

As for the regression case (chapter III), the asymptotic properties of the empirical distribution function F_n will depend on the mixing hypothesis on the process $(X_i)_N$ through sequences defined in the following manner.

When $(X_i)_N$ is φ-mixing (see definition 2.2.1), we introduce an increasing sequence of integers $(m_n)_N$ such that for any integer n we have

$$1 \leq m_n \leq n \text{ and } n\varphi_{m_n}(m_n)^{-1} \leq A < \infty, \tag{5.2.3}$$

where $(\varphi_n)_N$ are the mixing coefficients.

If the process $(X_i)_N$ is ρ-mixing (see definition 2.2.2), $(d_n)_N$ and $(q_n)_N$ are sequences defined from the mixing coefficients $(\rho_n)_N$ by

$$d_n = 8000 \exp\{3 \sum_{i=1}^{[\log n]} \rho_n^*(2^i)\}, \tag{5.2.4}$$

with

$$\rho_n^*(j) = \max_{1\leq k\leq n-j-1} \; \max_{1\leq m\leq (n-k-j)/2} |E\langle \sum_{i=1}^{k+m} X_i, \sum_{i=k+m+j}^{k+2m+j} X_i\rangle|,$$

and

$$n/q_n \geq 2 \text{ and } \limsup \, [2nq_n^{-1}]^{1/2}\rho(q_n) < 4/9. \tag{5.2.5}$$

When $(X_i)_N$ is α-mixing (see definition 2.2.3), we introduce a sequence $(s_n)_N$ which is an increasing sequence of integers such that for any integer n we have

$$1 \leq s_n \leq n/2 \text{ and } n\alpha_{s_n}^{(2s_n/3n)}/s_n \leq A < \infty, \qquad (5.2.6)$$

where $(\alpha_n)_N$ are the mixing coefficients.

We give in the theorem below the uniform convergence of F_n for mixing variables.

Theorem 5.2.1. Uniform convergence in φ-mixing case. (Collomb et al., 1985). If $(X_i)_N$ is φ-mixing and if

$$n/(m_n \log n) \xrightarrow[n\to\infty]{} \infty, \qquad (5.2.7)$$

then we have

$$\sup_{x\in G} |F_n(x)-F(x)| \xrightarrow[n\to\infty]{co.} 0. \qquad (5.2.8)$$

Theorem 5.2.2. Uniform convergence in ρ-mixing case. If $(X_i)_N$ is ρ-mixing and if

$$n/(d_n \log^2 n) \xrightarrow[n\to\infty]{} \infty \text{ and } n/(q_n \log n) \xrightarrow[n\to\infty]{} \infty, \qquad (5.2.9)$$

then we have

$$\sup_{x\in G} |F_n(x)-F(x)| \xrightarrow[n\to\infty]{co.} 0. \qquad (5.2.8)$$

Theorem 5.2.3. Uniform convergence in α-mixing case. If $(X_i)_N$ is α-mixing and if

$$n/(s_n \log n) \xrightarrow[n\to\infty]{} \infty, \qquad (5.2.10)$$

then we have

$$\sup_{x\in G} |F_n(x)-F(x)| \xrightarrow[n\to\infty]{co.} 0. \qquad (5.2.8)$$

Proof of theorems 5.2.1-5.2.3. We just sketch the proofs of these results since they are indeed very close to those of theorems 5.2.4-5.5.6 below. By an application of Collomb's inequality (see theorem 2.2.1) for φ-mixing variables (resp. theorem 2.2.2 for ρ-mixing variables), (resp. theorem 2.2.6 for α-mixing variables), we get

$$|F_n(x)-F(x)| \xrightarrow[n\to\infty]{co.} 0 \text{ for any } x \text{ in } G.$$

Then, in the three cases, using the increasing property of F and F_n and compactness of G we get (5.2.8).

We now give in the following theorems extension of Glivenko-Cantelli's theorem for mixing variables.

Theorem 5.2.4. Rate of convergence. Case of φ-mixing variables. (Sarda and Vieu, 1988b). If $(X_i)_N$ is φ-mixing, and if the sequence

$$\mu_n = (n^{-1} m_n \log n)^{1/2} \tag{5.2.11}$$

is such that

$$\lim_{n\to\infty} \mu_n = 0,$$

then we have

$$\mu_n^{-1} \sup_{x\in G} |F_n(x) - F(x)| = O(1), \text{ co.} \tag{5.2.12}$$

Proof of theorem 5.2.4. The proof consists in two steps. We first show that, using theorem 2.2.1, for any sequence $(\epsilon_n)_N$ such that $0 \leq \epsilon_n \leq B < \infty$, the following inequality holds

$$\forall n \geq n_0, \; \sup_{x\in G} P(|F_n(x)-F(x)| \geq \epsilon_n) \leq C_1 \exp(-C_2 n \epsilon_n^2 m_n^{-1}), \tag{5.2.13}$$

where C_1 and C_2 are positive constants independent of x. We thus define the sequence $(Z_i)_N$ as

$$Z_i = n^{-1}(I_{[X_i \leq x]} - F(x)).$$

We first remark that

$$F_n(x) - F(x) = \sum_{i=1}^{n} Z_i,$$

and that

$$EZ_i = 0, \; |Z_i| \leq 2n^{-1}, \; E|Z_i| \leq 2n^{-1}$$

and

$$E(Z_i^2) = n^{-2}F(x)(1-F(x)) \leq n^{-2}.$$

Then, applying Collomb's inequality (i.e. theorem 2.2.1), with $d=\delta=2n^{-1}$, $D=n^{-2}$ and $\alpha=n\epsilon_n/(8mB)$ for each integer m (depending on n) such that $m\leq n$, we get

$$P(|F_n(x)-F(x)|\geq\epsilon_n) \leq C_m \exp(-t(\epsilon_n,m)nm^{-1}), \tag{5.2.14}$$

where

$$t(\epsilon_n,m) = \epsilon_n^2 (8B)^{-1}(1-(8B)^{-1}(m^{-1}+16m^{-1}\bar{\varphi}_m)),$$

$$\bar{\varphi}_m = \sum_{i=1}^{m} \varphi_i,$$

and

$$C_m = 2\exp(3e^{1/2}nm^{-1}\bar{\varphi}_m).$$

We consider now the two possibilities for the sequence $(m_n)_N$.

First case: "$m_n \xrightarrow[n\to\infty]{} \infty$". By conditions (5.2.3) and $\varphi_n \xrightarrow[n\to\infty]{} 0$, we get the existence of an integer n_0 such that for any n greater than n_0 we have

$$m_n^{-1} \leq 2B \text{ and } m_n^{-1}\bar{\varphi}_{m_n} \leq B/8, \tag{5.2.15}$$

so that

$$t(\epsilon_n,m_n) \geq (16B)^{-1}\epsilon_n^2. \tag{5.2.16}$$

Applying (5.2.14) to $m = m_n$ and using (5.2.15) and (5.2.16), we obtain (5.2.13) with $C_1 = 2\exp(3Ae^{1/2})$ and $C_2 = (16B)^{-1}$.

Second case: "$\exists n_1\in N,\ \exists m_0\in N,\ \forall n\geq n_1,\ m_n=m_0$".
The inequalities (5.2.15) and (5.2.16) remain valid changing m_n in m_1, where m_1 is an integer independent of n and greater than m_0. Similarly (5.2.13) follows by applying (5.2.14) to $m = m_1$, with $C_1 = 2\exp(3Ae^{1/2})$ and $C_2 = (16B)^{-1}m_0m_1^{-1}$.

Now, in a second step, using the compactness property of G, we cover it, for any $p>0$, by a finite number k_p of balls of radius $p/2$,

$$G \subset \bigcup_{r=1}^{k_p} B_r, \tag{5.2.17}$$

where

$$B_r = \{(y_j)_{j=1}^d \in \mathbb{R}^d, \max_{j=1,\dots,d} |x_r^j - y_r^j| \le p/2\},$$

$x_r=(x_r^j)_{j=1}^d$ being some point of G. Let us define, for any $r=1,\dots,k_p$,

$$s_r = (s_r^j)_{j=1}^d \text{ with } s_r^j = x_r^j - p/2,$$

and

$$t_r = (t_r^j)_{j=1}^d \text{ with } t_r^j = x_r^j + p/2,$$

and note that

$$s_r \in B_r,\ t_r \in B_r, \text{ and } \forall x \in B_r,\ s_r \le x \le t_r.$$

For $r=1,\dots,k_p$, we have

$$F(s_r) \le \sup_{x\in B_r} F(x) \le F(t_r) \text{ and } F(s_r) \le \sup_{x\in B_r} F_n(x) \le F_n(t_r). \tag{5.2.18}$$

We have, using the differentiability property of F and the equivalence of the norms on $\mathbb{R}^d$,

$$|F(u) - F(v)| \le C \max_{j=1,\dots,d} |u^j - v^j| \text{ for any u and v in } \mathbb{R}^d.$$

Then we get by (5.2.17) and (5.2.18),

$$\sup_{x\in G} |F_n(x) - F(x)| \le \max_{r=1,\dots,k_p} \max_{z_r=s_r,t_r} |F_n(z_r) - F(z_r)| + Cd.$$

For fixed ϵ_0, and choosing $p \le \epsilon_0 \mu_n/(2C)$, the last inequality leads to

$$P(\mu_n^{-1} \sup_{x\in G} |F_n(x)-F(x)| \ge \epsilon_0)$$
$$\le \sum_{r=1}^{k_p} \sum_{z_r=s_r,t_r} P(\mu_n^{-1} |F_n(z_r)-F(z_r)| \ge \epsilon_0/2).$$

It suffices now to apply (5.2.13) to $\epsilon_n = \mu_n \epsilon_0$, with $\epsilon_0 > 2B^{-1/2}$ to complete the proof of theorem 5.2.4.

Theorem 5.2.5. Rate of convergence. Case of ρ-mixing variables. If $(X_i)_N$ is ρ-mixing and if the sequence

$$\mu_n = (n^{-1} d_n \log^2 n)^{1/2} + (n^{-1} q_n \log n), \tag{5.2.19}$$

is such that

$$\lim_{n\to\infty} \mu_n = 0,$$

we have

$$\mu_n^{-1} \sup_{x \in G} |F_n(x) - F(x)| = O(1) \text{ co.} \tag{5.2.20}$$

Proof of theorem 5.2.5. We follow the same developments as in theorem 5.2.4. Let us define

$$Z_i = n^{-1} (I_{[X_i \leq x]} - F(x)).$$

We apply the Bernstein-type inequality for ρ-mixing variables given in theorem 2.2.2, with

$$D = (d_n n^{-1})^{1/2}, \; d = 2n^{-1/2} d_n^{-1/2} (d_n n^{-1})^{1/2}.$$

Thus we have

$$E \exp(\alpha | \sum_{i=1}^{n} Z_i | / (d_n n^{-1})^{1/2}) \leq C,$$

where $\alpha = (2B)^{-1} c_1$, $B = c_2 + 8 q_n n^{-1/2} d_n^{-1/2}$, C, c_1 and c_2 are positives constants not depending on n. This last inequality leads to

$$P(| \sum_{i=1}^{n} Z_i | \geq \epsilon_n (d_n n^{-1})^{-1/2}) \leq C\exp(-\epsilon_n (d_n n^{-1})^{-1/2} \alpha)$$
$$\leq C\exp(-\epsilon_n c_1 / (c_2 d_n^{1/2} n^{-1/2} + 8 q_n n^{-1})).$$

From this inequality we complete the proof of theorem 5.2.5 following exactly the same arguments as those used to finish the proof of theorem 5.2.4 from inequality (5.2.13).

<u>Theorem 5.2.6.</u> **Rate of convergence. Case of α-mixing variables.** (Sarda and Vieu, 1989b). If $(X_i)_N$ is α-mixing, and if the sequence

$$\mu_n = (n^{-1}s_n \log n)^{1/2} \qquad (5.2.21)$$

is such that s_n satisfies (5.2.6) and

$$\lim_{n\to\infty} \mu_n = 0,$$

then we have

$$\mu_n^{-1} \sup_{x\in G} |F_n(x) - F(x)| = O(1), \text{ co.} \qquad (5.2.22)$$

<u>Proof of theorem 5.2.6.</u> As for theorem 5.2.4 it will be enough to show that for any sequence $(\epsilon_n)_N$ such that $0\leq\epsilon_n\leq B<\infty$, the following inequality holds

$$\forall n\geq n_0, \sup_{x\in G} P(|F_n(x)-F(x)|\geq\epsilon_n)\leq C_1\exp(-C_2 n\epsilon_n^2 s_n^{-1}), \qquad (5.2.23)$$

where C_1 and C_2 are positive constants independent of x, since then we will get theorem 5.2.6 from this inequality exactly as we got theorem 5.2.4 from inequality (5.2.13). To show (5.2.3) we want to use the exponential inequality given in theorem 2.2.6. For this, we thus define the sequence $(Z_i)_N$ as

$$Z_i = n^{-1}(I_{[X_i\leq x]} - F(x)).$$

We first remark that

$$F_n(x) - F(x) = \sum_{i=1}^{n} Z_i,$$

and that

$$EZ_i = 0, \ |Z_i| \leq 2n^{-1},$$

and

$$E(Z_i^2) = n^{-2}F(x)(1-F(x)) \leq n^{-2}.$$

Then, applying theorem 2.2.6, we get (using the notations introduced in this theorem)

$$P(|F_n(x)-F(x)|\geq\epsilon_n) \leq C_k\exp(-\epsilon_n^2 nk^{-1}t(k)), \qquad (5.2.24)$$

where

$$t(k) = (8eB)^{-1} \, (1-(2B)^{-1}(k^{-1}+32k^{-1}\sum_{i=1}^{k}\alpha_i)),$$

and

$$C_k = 2\exp(3e^{1/2}nk^{-1}\alpha_k^{2k/3n}).$$

We consider now the two possibilities for the sequence $(s_n)_N$.

First case: "$s_n \xrightarrow[n\to\infty]{} \infty$". In this case there exists an integer n_0 such that for any n greater than n_0 we have

$$s_n^{-1} \leq 2B \text{ and } s_n^{-1}\sum_{i=1}^{s_n}\alpha_i \leq B/8, \qquad (5.2.25)$$

so that

$$t(s_n) \geq (16Be)^{-1}. \qquad (5.2.26)$$

Applying (5.2.24) to $k = s_n$ and using (5.2.25) and (5.2.26), we obtain (5.2.23) with $C_1 = 3\exp(2Ae^{1/2})$ and $C_2 = (16Be)^{-1}$.

Second case: "$\exists n_1\in N,\ \exists s_0\in N,\ \forall n\geq n_1,\ s_n=s_0$". The inequalities (5.2.25) and (5.2.26) remain valid changing s_n in s_1, where s_1 is an integer independent of n and greater than s_0. Similarly (5.2.23) follows by applying (5.2.24) to $k=s_1$, with $C_1=3\exp(2Ae^{1/2})$ and $C_2 = (16Be)^{-1}s_0s_1^{-1}$.

So, in both cases (5.2.23) is proven and the proof of theorem 5.2.6. is therefore complete.

Remark 5.2.1. We mention that the arguments involved in the proof of theorems 5.2.4-5.2.6 are quite different to those usually found in the literature concerning empirical distribution function under dependence. Many author dealing with this problem investigate techniques based on strong approximation of the empirical process limiting their study to the univariate case, and most often for uniformly distributed variables (see e.g. Berkes and Philipp, 1977 , Philipp, 1977 and references therein). The analytic argument used here allows us to deal with the general case. Nevertheless, we note that an improvement of the rate of convergence in (5.2.10) and (5.2.18) seems to be difficult to obtain by such a technique.

3. Uniform consistency of kernel density estimates.

Backed by the results of the above section 5.2 we need now uniform consistency results of density estimates, in order to give consistency results of hazard estimates of form (5.1.2). Let us look at the kernel density estimate f_n (Rosenblatt, 1956b and Parzen, 1962),

$$f_n(x) = (nh^d)^{-1} \sum_{i=1}^{n} K((x-X_i)/h,$$

where K is a $\mathbb{R}^d$ valued kernel function and where the bandwidth h (depending on n) is a positive real number. Note that the structure of the proofs of results of section 3.3. consists in treating separately the numerator from the denominator of the regression estimate r_n (see proof of theorem 3.2.2). And note also that the denominator of the regression estimate r_n was just the density estimate f_n. Therefore, the following results were stated along the proofs of results of section 3.3. In the following the sequences $(m_n)_N$, $(s_n)_N$, $(q_n)_N$ and $(d_n)_N$ are those defined by (5.2.3)-(5.26).

Theorem 5.3.1. **φ-mixing case.** Assume that f is k-times differentiable and that its derivatives of order k are Lipschitz continuous of order γ. Assume that (K.1)-(K.6) hold and that $(X_i)_N$ is φ-mixing. If the bandwidth is such that the sequence

$$V_n = h^{k+\delta} + [(m_n \log n)/(nh^d)]^{1/2}$$

satisfies the condition

$$V_n \longrightarrow 0 \text{ as } n \longrightarrow \infty,$$

then we have

$$\sup_{x\in G} |f_n(x) - f(x)| = O(V_n), \text{ co.}$$

This rate of convergence can be optimised by a suitable asymptotic choice of the bandwidth h.

Corollary 5.3.1. Assume that the conditions of theorem 5.3.1 hold and that the bandwidth is taken to be, for some real positive constant C, of the form

$$h^* = C\,(n^{-1} m_n \log n)^{1/(2k+2\gamma+d)},$$

then we obtain a rate of convergence in

$$V_n^* = (n^{-1} m_n \log n)^{(k+\gamma)/(2k+2\gamma+d)}.$$

We have the same kind of results when the variables are ρ-mixing.

Theorem 5.3.2. **ρ-mixing case.** Assume that f is k-times differentiable and that its derivatives of order k are Lipschitz continuous of order γ. Assume that (K.1)-(K.6) hold and that $(X_i)_N$ is ρ-mixing. If the bandwidth is such that the sequence

$$W_n = h^{k+\gamma} + [(d_n \log^2 n)/(nh^d)]^{1/2} + (q_n \log n)/(nh^d),$$

satisfies the condition

$$W_n \longrightarrow 0,$$

then we have

$$\sup_{x \in G} |f_n(x) - f(x)| = O(W_n), \text{ co.}$$

As before, choosing h balancing the trade-off between the different terms of W_n, allows us to optimise these rates of convergence in some particular cases.

Corollary 5.3.2. Assume that the conditions of theorem 5.3.2 hold and that we have $\rho_n = O(a^{-n})$ for some $a > 1$. If the bandwidth is taken to be, for some finite positive constant C, of the form

$$h^* = C\,(n^{-1} \log^2 n)^{1/(2k+2\gamma+d)},$$

then we obtain a rate of convergence in

$$W_n^* = (n^{-1} \log^2 n)^{(k+\gamma)/(2k+2\gamma+d)}.$$

Corollary 5.3.3. Assume that the conditions of theorem 5.3.2 hold and that we have $\rho_n \leq l n^{-a}$ for some $a > 0$ and $l > 0$.

1st case: $a < (d+k+\gamma)/(2k+2\gamma)$. If the bandwidth is taken to be, for some finite positive constant C, of the form

$$h^* = C(n^{-2a/(2a+1)} \log n)^{1/(d+k+\gamma)},$$

then we get a rate of convergence in

$$W_n^* = (n^{-2a/(2a+1)} \log n)^{(k+\gamma)/(2k+2\gamma+d)}.$$

2nd case: $a > (d+k+\gamma)/(2k+2\gamma)$. If the bandwidth is taken to be, for some finite positive constant C, of the form

$$h^* = C\,(n^{-1}\log^2 n)^{1/(2k+2\gamma+d)},$$

then we obtain a rate of convergence in

$$W_n^* = (n^{-1}\log^2 n)^{(k+\gamma)/(2k+2\gamma+d)}.$$

We have similar results when the dependence structure between the variables is the α-mixing condition.

Theorem 5.3.3. **α-mixing case.** Assume that f is k-times differentiable and that its derivatives of order k are Lipschitz continuous of order γ. Assume that (K.1)-(K.6) hold and that $(X_i)_N$ is α-mixing. If the bandwidth is such that the sequence

$$S_n = h^{k+\alpha} + [(s_n \log n)/(nh^d)]^{1/2}$$

satisfies the condition

$$S_n \longrightarrow 0 \text{ as } n \longrightarrow \infty,$$

then we have

$$\sup_{x \in G} |f_n(x) - f(x)| = O(S_n), \text{ co.}$$

This rate of convergence can be optimised by a suitable asymptotic choice of the bandwidth h.

Corollary 5.3.4. Assume that the conditions of theorem 5.3.3 hold and that the bandwidth is taken to be, for some real positive constant C, of the form

$$h^* = C\,(n^{-1}s_n \log n)^{1/(2k+2\gamma+d)},$$

then we obtain a rate of convergence in

$$S_n^* = (n^{-1}s_n \log n)^{(k+\gamma)/(2k+2\gamma+d)}.$$

4. Kernel hazard estimation.

4.1. Definition and assumptions.

The kernel hazard estimate g_n was first introduced by Murthy (1965). It is of the form (5.1.2) with f_n being the so-called kernel density estimate (Rosenblatt, 1956b and Parzen, 1962) given, from a density function K on $\mathbb{R}^d$, the kernel, and a real number h (depending on n), referred as the smoothing parameter or the bandwidth, by

$$f_n(x) = (nh^d)^{-1} \sum_{i=1}^{n} K((x-X_i)/h) \text{ for any } x \text{ in } \mathbb{R}^d, \quad (5.4.1)$$

and F_n is the empirical distribution function.

In the case d = 1 and for independent variables, Murthy (1965) proves the almost sure pointwise convergence of g_n together with asymptotic normality of this estimate.

As seen in previous sections, the reason to deal with this estimate is based on practical aspects, its easy computation as well as the theoretical tractability. Furthermore it achieves in the independent case the optimal rate of convergence for an estimate of g based on a ratio of an estimate of f over an estimate of 1-F. This result is derived by applying (5.1.3) and the fact that f_n reaches (for independent variables) the optimal rate given by Stone (1983) (see Sarda and Vieu, 1988b) for h convenient and that the rate of convergence for F_n is of smaller order.

The compact G is defined as above and we suppose moreover that f is bounded on $\mathbb{R}^d$ and bounded away from zero on $\hat{G}$:

$$\exists\Gamma<\infty,\ f(x) \leq \Gamma,\ \forall x\in\mathbb{R}^d, \quad (5.4.2)$$

$$\exists\tau>0,\ f(x) \geq \tau,\ \forall x\in\hat{G}. \quad (5.4.3)$$

The kernel function K is assumed to be a real function, bounded, belonging to $L^2(\mathbb{R}^d)$ such that

$$|z|^d k(z) \xrightarrow[|z|\to\infty]{} 0,$$

and

$$\int K(z)dz = 1.$$

Moreover we suppose that the kernel K satisfy the following Lipschitz condition

$$\exists\gamma>0,\ \forall(z,z')\in\mathbb{R}^{2d},\ |K(z)-K(z')| \leq M\|z-z'\|^\gamma,\ M<\infty. \quad (5.4.4)$$

We will see in next section that this last assumption is assumed to make the proofs simplier and that other kernels can be introduced in some cases. Nevertheless, we note that condition (5.4.4) is commonly assumed in literature about kernel estimation.

The rate of convergence of g_n will be given in function of the number k of vanishing moments of the kernel function K: we suppose that k is a positive integer such that

$$\int \| z \|^j K(z)dz = 0 \text{ for } j = 1,\dots,k$$

and (5.4.5)

$$0 < \left|\int \| z \|^{k+1} K(z)dz\right| < \infty.$$

Such kernels, referred as kernel of order k in the literature, are introduced for example by Gasser and Müller (1979) in order to minimize bias in kernel regression estimation (see also Lejeune and Sarda, 1988).

4.2. Case of φ-mixing variables.

We assume here that the variables $(X_i)_N$ are φ-mixing. The sequence $(m_n)_N$ is defined as in previous sections and verifies with the coefficients $(\varphi_n)_N$ the condition (5.2.3). We give in a first theorem uniform convergence of the estimate g_n under this mixing condition.

Theorem 5.4.1. Uniform convergence. (Collomb et al., 1985). Suppose that conditions (5.4.2)-(5.4.4) hold. Moreover, if the density function f is continuous on $\mathbb{R}^d$ and if the bandwidth is such that

$$nh^d/(m_n \log n) \xrightarrow[n\to\infty]{} \infty, \tag{5.4.6}$$

then we have

$$\sup_{x\in G} |g_n(x)-g(x)| \xrightarrow[n\to\infty]{co.} 0. \tag{5.4.7}$$

Proof of theorem 5.4.1. This theorem is just an obvious consequence of theorem 5.4.2 below.

Remark 5.4.1. For Markovian processes with Doeblin's condition (see section III.3.4), the condition (5.4.6) becomes

$$nh^d/(\log n)^2 \xrightarrow[n\to\infty]{} \infty,$$

and for m-dependent variables (and therefore for independent ones) it becomes

$$nh^d/\log n \xrightarrow[n\to\infty]{} \infty.$$

Remark 5.4.2. The result (5.4.7) holds for a kernel of the form

$$K_B(z) = I_B(z),\ \forall z\in\mathbb{R}^d, \tag{5.4.10}$$

where B is a compact of $\mathbb{R}^d$, with $0\in B$, of Lebesgue's measure 1 and such that

$$\forall z\in B,\ \forall c\in[0,1],\ cz\in B.$$

Proof of remark 5.4.2. The proof consists in applying Urysohn's lemma which allows us to define for any $\eta\in]0,1[$, two kernels k^η and K^η verifying the Lipschitz condition (5.4.4) and such that

$$k^\eta(z) \le K_B(z) \le K^\eta(z),\ \forall z\in\mathbb{R}^d,$$

and (5.4.11)

$$|k^\eta - K^\eta| \le \eta \text{ and } |K^\eta - K_B| \le \eta.$$

We get

$$f_n^{k^\eta}(x) - Ef_n^{k^\eta}(x) + \alpha_n^{k^\eta}(x) \le f_n^{K_B}(x) - Ef_n^{K_B}(x) \le f_n^{K^\eta}(x) - Ef_n^{K^\eta}(x) + \alpha_n^{K^\eta}(x),$$

where

$$\alpha_n^K(x) = Ef_n^K(x) - Ef_n^K(x) \text{ for } K = k^\eta \text{ or } K^\eta.$$

Using (5.4.2) and (5.4.11) we get by integration

$$\alpha_n^K \le \Gamma\eta,$$

and then we have

$$\sup_{x\in G} |f_n^{K_B}(x) - Ef_n^{K_B}(x)| \le \Gamma\eta + \max_{K=k^\eta,K^\eta} \sup_{x\in G} |f_n^K(x) - Ef_n^K(x)|.$$

Then by (5.4.8) we obtain

$$\sup_{x\in G} |f_n^{K_B}(x) - Ef_n^{K_B}(x)| \xrightarrow[n\to\infty]{co.} 0.$$

The result (5.4.9) is derived directly as for a Lipschitzian kernel, and (5.4.7) follows by using theorem 5.2.2. This completes the proof of remark 5.4.2

The rate of convergence in (5.4.7) is specified in the following result.

Theorem 5.4.2. Rate of convergence of the kernel hazard estimate. (Sarda and Vieu, 1988b). Under the conditions (5.4.2)-(5.4.5), if the density f is k-times differentiable and its derivatives of order k satisfies a Lipschitz condition of order γ and the sequence

$$V_n = h_n^{k+\gamma} + (m_n \log n)^{1/2} (nh^d)^{-1/2}, \tag{5.4.12}$$

is such that

$$\lim_{n\to\infty} V_n = 0,$$

we have

$$V_n^{-1} \sup_{x\in G} |g_n(x)-g(x)| = O(1) \text{ co.} \tag{5.4.13}$$

__Proof of theorem 5.4.2.__ We obtain (5.4.13) by direct application of the second part of (5.1.3), theorems 5.2.4 and 5.3.1.

By taking a suitable asymptotic smoothing parameter we can optimise the rate of convergence of the hazard kernel estimate.

__Corollary 5.4.1.__ Under conditions of theorem 5.4.2 and for a bandwidth h^* of the form,

$$h^* = C(n^{-1} m_n \log n)^{\frac{1}{(2k+2\gamma+d)}},$$

C being a positive real constant, we have

$$\sup_{x\in G} |g_n(x)-g(x)| = O((n^{-1} m_n \log n)^{\frac{k+\gamma}{(2k+2\gamma+d)}}) \text{ co.} \tag{5.4.14}$$

__Remark 5.4.3.__ In the particular case of independent variables $(X_n)_N$, choosing $m_n=1$, $\forall n\in N$, the rate of convergence is

$$(n^{-1} \log n)^{\frac{k+\gamma}{(2k+2\gamma+d)}}, \tag{5.4.15}$$

and when the process is of Markovian stucture, choosing m_n to be proportional to logn, the rate becomes

$$(n^{-1} \log^2 n)^{\frac{k+\gamma}{(2k+2\gamma+d)}}.$$

__Remark 5.4.4.__ As pointed out in section V.3.1, the rate of convergence of g_n depends only on the rate of convergence of f_n. Therefore, since the rate of convergence obtained in (5.4.13) is optimal for the estimation of density function (see Farell, 1982 and Stone, 1983), it is also optimal for the estimation of the hazard function in the class of estimates based on a ratio of an estimate of f over an estimate of F. This last assertion is proved via inequalities (5.1.3).

4.3. Case of ρ-mixing variables.

We assume now in this section that the variables $(X_i)_N$ are ρ-mixing (see definition 2.2.2). The sequences $(d_n)_N$ and $(q_n)_N$ are defined as in section 5.2 (formulas (5.2.4) and (5.2.5)).

Theorem 5.4.3. Uniform convergence. Suppose that conditions (5.4.2)-(5.4.4) hold. If the density function f is continuous on G and if the bandwidth is such that

$$nh^d/(d_n \log^2 n) \longrightarrow \infty \text{ and } nh^d/(q_n \log n) \longrightarrow \infty, \qquad (5.4.16)$$

then we have

$$\sup_{x \in G} |g_n(x)-g(x)| \xrightarrow[n\to\infty]{co.} 0. \qquad (5.4.17)$$

Proof of theorem 5.4.3. The technique of the proof is similar to that of theorem 5.3.1 below; it consists in showing (using (5.1.3)) equivalents of (5.4.17) for f_n and F_n under ρ-mixing condition. In fact, theorem 5.4.3 is just an obvious corollary of theorem 5.4.4 below.

Remark 5.4.5. The result (5.4.17) can be extended for a kernel of the form (5.4.10).

Proof of remark 5.4.5. The proof is derived involving the same argument as in remark 5.3.2.

We precise now the rate of convergence in (5.4.17) in the following theorem.

Theorem 5.4.4. Rate of convergence. Under the conditions (5.4.2)-(5.4.5), if the density f is k-times differentiable and its derivative of order k satisfies a Lipschitz condition of order γ and the sequence

$$W_n = h^{k+\gamma}+(d_n \log^2 n)^{1/2}(nh^d)^{-1/2}+(q_n \log n)/(nh^d), \qquad (5.4.18)$$

is such that

$$\lim_{n\to\infty} W_n = 0,$$

we have

$$W_n^{-1} \sup_{x \in G} |g_n(x)-g(x)| = O(1) \text{ co.} \qquad (5.4.19)$$

Proof of theorem 5.4.4. The proof results from (5.1.3) and theorem 5.2.5 and theorem 5.3.2.

As pointed out in section 3.7.2 it is not possible to optimise the rate of convergence in W_n in the general case. Nevertheless we have the following results.

Corollary 5.4.2. Under the conditions of theorem 5.4.6, if the mixing coefficients verify $\rho_n = O(a^{-n})$ for a<1, and for a bandwidth h^* such that, for some finite positive constant C,

$$h^* = C(n^{-1}\log^2 n)^{1/(2k+2\gamma+d)},$$

we have

$$\sup_{x\in G} |g_n(x)-g(x)| = O((n^{-1}\log^2 n)^{k+\gamma/(2k+2\gamma+d)}). \quad (5.4.20)$$

Corollary 5.4.3. Under the conditions of theorem 5.4.6 and if the mixing coefficients verify $\rho_n \leq ln^{-a}$ for a>0 and l>0, we have

i) if $a<(d+k+\gamma)/(2k+\gamma)$ and for the bandwidth h^* such that

$$h^* = C(n^{-2a/(2a+1)}\log n)^{1/(d+k+\gamma)},$$

C being a real positive constant, we have

$$\sup_{x\in G}|g_n(x)-g(x)|=O(n^{-2a/(2a+1)}\log n)^{(k+\gamma)/(2k+2\gamma+d)}, \text{ co.} \quad (5.4.21)$$

ii) if $a> (d+k+\gamma)/(2k+2\gamma+d)$ and for the bandwidth h^* such that

$$h_n^* = O((n^{-1}\log^2 n)^{1/(2k+2\gamma+d)}),$$

C being a finite real positive constant, we have

$$\sup_{x\in G}|g_n(x)-g(x)|=O(n^{-1}\log^2 n)^{(k+\gamma)/(2k+2\gamma+d)}), \text{ co.} \quad (5.4.22)$$

4.4. Case of α-mixing variables.

We assume here that the variables $(X_i)_N$ are α-mixing. The sequence $(s_n)_N$ is defined as in previous sections and verifies with the coefficients $(\alpha_n)_N$ the condition (5.2.6). We give in a first theorem uniform convergence of the estimate g_n under this mixing condition.

Theorem 5.4.7. Uniform convergence. Suppose that conditions (5.4.2)-(5.4.4) hold. Moreover, if the density function f is continuous on $\mathbb{R}^d$ and if the bandwidth is such that

$$nh^d/(s_n \log n) \xrightarrow[n\to\infty]{} \infty, \tag{5.4.23}$$

then we have

$$\sup_{x\in G} |g_n(x)-g(x)| \xrightarrow[n\to\infty]{co.} 0. \tag{5.4.24}$$

Proof of theorem 5.4.7. This result is a trivial consequence of theorem 5.4.8 below.

Remark 5.4.6. The result (5.4.24) holds for a kernel of the form of (5.4.10).

Proof of remark 5.4.6. This proof performs exactly as the proof of remark 5.4.2 above.

The rate of convergence in (5.4.24) is specified in the following result.

Theorem 5.4.8. Rate of convergence. (Sarda and Vieu, 1989a). Under the conditions (5.4.2)-(5.4.5), if the density f is k-times differentiable and its derivatives of order k satisfies a Lipschitz condition of order γ and the sequence

$$S_n = h_n^{k+\gamma} + (s_n \log n)^{1/2} (nh^d)^{-1/2}, \tag{5.4.25}$$

is such that

$$\lim_{n\to\infty} S_n = 0,$$

then we have

$$S_n^{-1} \sup_{x\in G} |g_n(x)-g(x)| = O(1) \text{ co.} \tag{5.4.26}$$

Proof of theorem 5.4.8. We obtain (5.4.13) by direct application of the second part of (5.1.3), theorems 5.2.6 and 5.3.3.

By taking a suitable asymptotic smoothing parameter we can optimise the rate of convergence of the hazard kernel estimate.

Corollary 5.4.4. Under conditions of theorem 5.4.8 and for a bandwidth h^* such that, for some finite positive constant C,

$$h^* = O(n^{-1}s_n \log n)^{\frac{1}{(2k+2\gamma+d)}},$$

we have

$$\sup_{x\in G} |g_n(x)-g(x)| = O((n^{-1}s_n \log n)^{\frac{k+\gamma}{(2k+2\gamma+d)}}) \text{ co.} \qquad (5.4.27)$$

5. k-NN hazard estimation.

5.1 Definition and assumptions.

The k-nearest neighbour hazard estimate $\hat{g}_n$ is defined by (5.1.2) with f_n replaced by $\hat{f}_n$, $\hat{f}_n$ being the k-NN density estimate. This density estimate is defined, from a kernel K on $\mathbb{R}^d$, by

$$\hat{f}_n(x) = (nR(k_n,x)^d)^{-1} \sum_{i=1}^{n} K\left(\frac{x-X_i}{R(k_n,x)}\right), \qquad (5.5.1)$$

for any x in $\mathbb{R}^d$, where

$$R(k_n,x) = \inf \{a\in\mathbb{R}^+,\ \#\{X_i,\ i=1,\dots,n,\ |x-X_i|\le a\}\ge k_n\},$$

and $(k_n)_N$ is a sequence of positive integers such that

$$k_n/n \xrightarrow[n\to\infty]{} 0.$$

This estimate was first introduced by Loftsgaarden and Quesenberry (1965).

The kernel function K is defined as in section 5.4.1, verifies 5.4.4 (resp. 5.4.10) and moreover we suppose that it is such that

$$K(cz) \ge K(z),\ \forall c\in[0,1],\ \forall z\in\mathbb{R}^d \qquad (5.5.2a)$$

and

$$K(z) = 0,\ \forall z\in\mathbb{R}^d,\ |z| > 1. \qquad (5.5.2b)$$

The compact G is defined as above, i.e. it is a compact subset of the support of 1-F, and $\hat{G}$ is a compact ϵ-neighborhood of G in $\mathbb{R}^d$ on which f is bounded away from zero.

The main tool in proving consistency results for k-NN estimate is the following lemma. This lemma allows, under general assumptions, to derive consistency results for k-NN estimates directly from similar results for kernel estimates.

Lemma 5.5.1. Assume that K is a kernel function which verifies (5.5.2). Every convergence property (pointwise or uniform, almost surely or completely) of the kernel density estimate f_n defined in (5.4.1) with kernel K and with bandwidth h, remains valid for the k-NN density estimate $\hat{f}_n$ defined with the same kernel K and with $k_n \approx nh^d$.

Proof of lemma 5.5.1. This lemma was proven by Moore and Yackel (1977, theorem 1.1) when the variables X_i were independent. But, as pointed out by Collomb et al. (1985), their proof did not use this independence assumption and is therefore still valid for dependent X_i. So, we refer to Moore and Yackell's paper for a proof of this lemma.

In the following, we will use this lemma to get, under several different mixing conditions, consistency results for the k-NN hazard estimates directly from results of sections 3 and 4 of this chapter.

5.2. Case of φ-mixing variables.

Let us assume that $(X_i)_N$ is φ-mixing with $(m_n)_N$ being a sequence of integer such that (5.2.3) holds. We have then the following theorem.

Theorem 5.5.1. Uniform convergence. (Collomb et al., 1985) Let K verifies (5.5.2) and (5.4.4) (resp. (5.4.10)). Then, if conditions (5.4.2)-(5.4.4) hold, if f is continuous on $\mathbb{R}^d$ and if $(k_n)_N$ is such that

$$k_n/(m_n \log n) \xrightarrow[n\to\infty]{} \infty, \tag{5.5.3}$$

then we have

$$\sup_{x\in G} |\hat{g}_n(x)-g(x)| \xrightarrow[n->\infty]{co.} 0. \tag{5.5.4}$$

Proof of theorem 5.5.1. Result (5.5.4) is derived immediately from theorem 5.4.1 and remark 5.4.2, together with the lemma 5.5.1 above.

5.3. Case of ρ-mixing variables.

Let us assume that $(X_i)_N$ is ρ-mixing and the sequences $(d_n)_N$ and $(q_n)_N$ are defined by (5.2.4) and (5.2.5).

Theorem 5.5.2. Uniform convergence. Let K verifies (5.5.2) and (5.4.4) (resp. (5.4.10)). If conditions (5.4.2) and (5.4.3) hold, if f is continuous on G and if the sequence $(k_n)_N$ is such that

$$k_n/(d_n \log^2 n) \longrightarrow \infty \text{ and } k_n/(q_n \log n) \longrightarrow \infty, \tag{5.5.5}$$

then we have

$$\sup |\hat{g}_n(x)-g(x)| \xrightarrow[n\to\infty]{co.} 0. \tag{5.5.6}$$

Proof of theorem 5.5.2. The proof follows from theorem 5.4.3 and lemma 5.5.1.

5.4. Case of α-mixing variables.

Let us assume that $(X_i)_N$ is α-mixing with $(s_n)_N$ being a sequence of integer such that (5.2.6) holds. We have then the following theorem.

Theorem 5.5.3. Uniform convergence. Let K verifies (5.5.2) and (5.4.4) (resp. (5.4.10)). Then if conditions (5.4.2)-(5.4.4) hold, if f is continuous on $\mathbb{R}^d$ and if the sequence $(k_n)_N$ is such that

$$k_n/(s_n \log n) \xrightarrow[n\to\infty]{} \infty,$$

then we have

$$\sup_{x\in G} |\hat{g}_n(x)-g(x)| \xrightarrow[n->\infty]{co.} 0.$$

Proof of theorem 5.5.3. It follows from theorem 5.4.5 together with lemma 5.5.1.

Chapter VI

HOW TO SELECT THE SMOOTHING PARAMETER?

1. The smoothness problem.

1.1. Introduction.

From the results given in the previous sections it appeared that the bandwidth h played a dominant role in the behaviour of kernel estimates for regression, density or hazard function estimation.

Let us look for instance at the kernel regression setting. The prominent role of the bandwidth h can be seen from corollaries 3.3.1-3.3.4. The rates of uniform convergence are basically the sum of two components, i.e bias and variance. Since bias (resp. variance) is proportional to h (resp. proportional to h^{-1}) the bandwidth has to be taken not too large (resp. too small) since it would increase the bias (resp. the variance) of the estimate. In the above mentioned corollaries, in order to optimise the rate of convergence we had to choose a bandwidth that balances this trade-off between bias and variance components.

This phenomenon is exactly the same in density estimation or in hazard estimation. In effect, results similar to corollaries 3.3.1-3.3.4 have been stated in section 3 (resp. in section 4) of chapter V for kernel density estimate (resp. for kernel hazard estimate).

1.2. An example.

Before giving theoretical results, and to make evident the importance of the bandwidth, let us look at the following simulated example. An autoregressive process has been generated in the following way:

$$X_{i+1} = R(X_i) + \epsilon_i, \quad i=1,\dots,500,$$

where ϵ_i are independent random variables simulated from an uniform distribution on the interval $[-1/2,+1/2]$. The autoregression function R was taken to be

$$R(x) = \begin{cases} x/(1+x^2) & \text{if } -1 \leq x \leq 1; \\ 0 & \text{otherwise.} \end{cases} \qquad (6.1.1)$$

This process is φ mixing by the results of chapter III. In order to apply a nonparametric kernel estimator to these simulated data we used as kernel function the so-called Epanechnikov kernel, i.e.,

$$K(x) = \begin{cases} 3/4\,(1-x^2) & \text{if } -1 \leq x \leq 1; \\ 0 & \text{otherwise.} \end{cases} \tag{6.1.2}$$

The effects of the bandwidth on the behavior of the estimate are shown in figures 6.1.1, 6.1.2 and 6.1.3.

Figure 6.1.1 shows the graph of the true function R defined by (6.1.1) together with the graph of the kernel predictor (see definition in section 3.4.2) which uses a bandwidth h=.10.

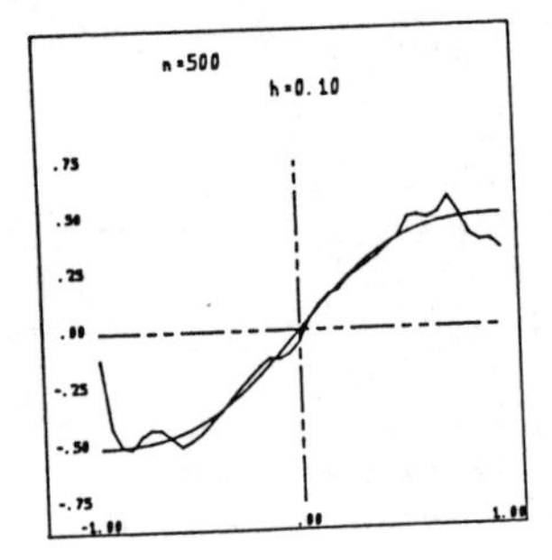

Figure 6.1.1
Undersmoothed Estimate

Clearly such an estimate is undersmoothed. The bandwidth used was too small and the estimate has important variance. The same curves are shown in figure 6.1.2 when the bandwidth is taken to be h=.34. In this case, the bandwidth used was too large and it resulted an oversmoothed estimation.

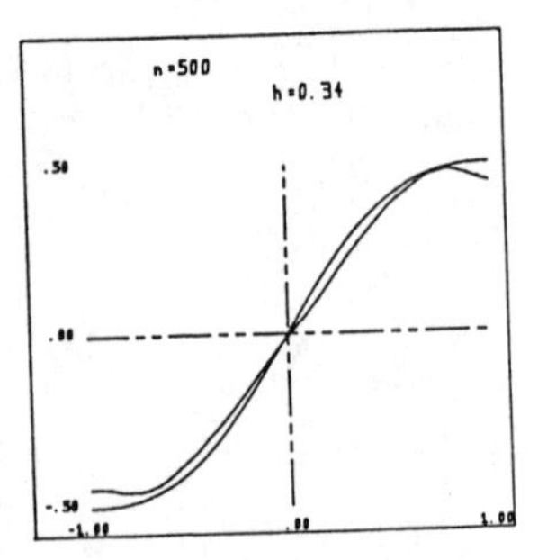

Figure 6.1.2
Oversmoothed Estimate

The corresponding estimate has high bias. Figure 6.1.3 shows what happens when the bandwidth is h=.22. This value for the smoothing parameter gives a proper estimation, since the corresponding estimate balances the trade-off between variance and bias effects (i.e. between undersmoothing and oversmoothing).

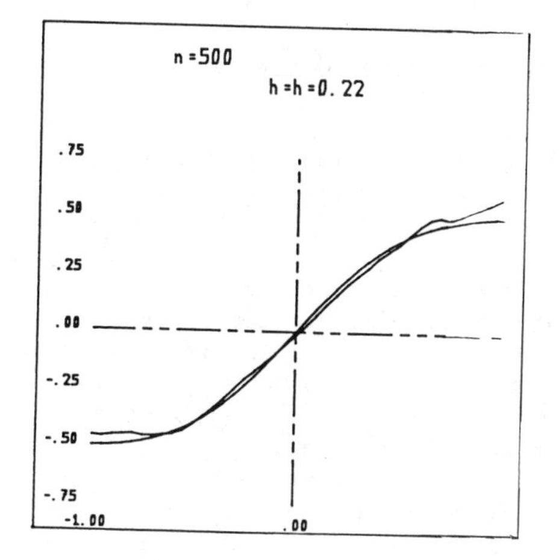

Figure 6.1.3
Optimal Estimation, h = 0.22

1.3. The amount of smoothness.

In practical situations it is certainly desirable to determine an appropriate bandwidth in some automatic fashion. As shown in theabove example, a bad choice for the smoothing parameter may lead to very poor estimates.

Unfortunately, this cannot be derived from results of the kind of corollaries 3.3.1 or 3.3.4 above. Such results give indications about the rate of convergence of the optimal bandwidth, but this rate is useless in practice since it depends on the number of derivatives of the unknown function to estimate. Moreover, even if we knew this number of derivatives, the constant term involved in the formulation of these optimal bandwidths includes quantities which are also depending on the unknown function.

So, the problem is to find a data-driven (i.e. depending only the data and then directly computable in practice) bandwidth, say $\hat{h}$, which is "as good as" the "best theoretical" bandwidth. A reasonable way to quantify this is the following definition which applies in all functional estimation problems (regression, density, hazard, ...). Let d( ) be a measure of accuracy for a kernel estimate (d is in fact a distance between the estimator and the function to estimate), and let it depend on the smoothing parameter h.

Definition 6.1.1. (Shibata, 1981). A data-driven smoothing parameter $\hat{h}$ is said asymptotically optimal with respect to d if we have

$$d(\hat{h})/\inf d(h) \xrightarrow{a.s.} 1.$$

1.4. The aim of cross-validation procedures.

One of the practical methods to solve this smoothness problem, is the cross-validation procedure. This procedure has been investigated by a lot of authors in the setting of independent variables (see, e.g., Hall, 1984, Härdle and Marron, 1985 and Härdle, Hall and Marron, 1988 in regression setting, Rudemo, 1982, Bowman, 1984, Hall, 1983, Stone, 1984, Marron, 1987, Hall and Marron, 1987 and Mielniczuk et al, 1988 in density estimation and Sarda and Vieu, 1989a in hazard estimation)). The aim of these cross-validation procedures is to find data-driven smoothing parameter that asymptotically minimises some quadratic measure of error. In the following sections we will explicitely describe these procedures and we will give asymptotically optimality results.

But, to make cross-validation more evident, let us turn for a short moment to the simulated example above. Cross- validation consists in introducing a data-driven criterion, let call it CV, which is approximating a theoretical quadratic error, let say for example averaged square error d_A, up to a constant term (i.e. a term independent of h). This can be seen from figure 6.1.3. which shows, for the simulated example described above, the curves of the cross-validation criterion CV and of the theoretical error d_A as functions of the bandwidth h.

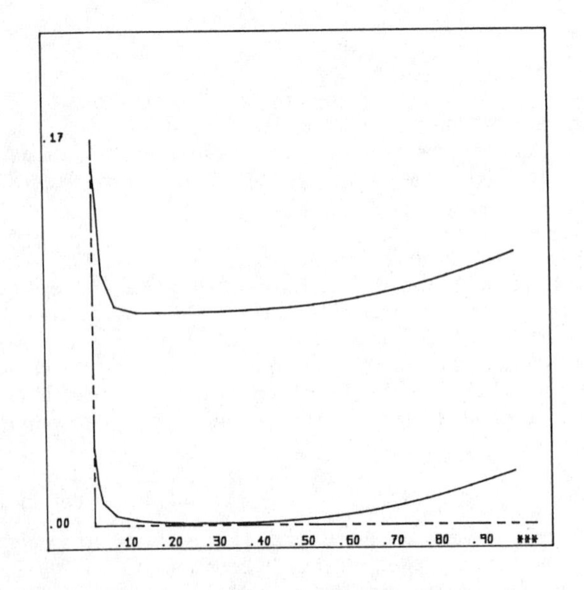

Figure 6.1.4
CV and d_A functions

On this example it can be seen that the data-driven cross- validation criterion and the theoretical error have roughly the same minimum since theoretical error DA and cross-validation criterion CV have roughly the same curve (up to a constant term). Therefore, the cross-validated bandwidth that will be obtained by taking $\hat{h}$ to be the minimiser of the data-driven criterion CV would be expected to be a suitable estimate of the optimal theoretical bandwidth, i.e. of the minimiser of DA. In fact we expect to have asymptotic optimality in the sense of definition 6.1.1 above for $d=d_A$.

In the following, such asymptotic optimality results will be stated for each of the three following problems :

-density estimation (section 3);

-hazard estimation (section 4);

-regression estimation (section 5).

These results will be given in the setting of α-mixing variables. This mixing condition is the less restrictive among those studied in this book. Before that, we need some mathematical tool that is more powerful than those necessary to get consistency results and used untill now along this book. This tool will be specified in section 2 below. The result of this section 2 can be seen as an extension, in a very particular case, of the inequalities given in section 2.4 of chapter II.

2. Main mathematical tool.

Asymptotic optimality properties are usually stated in the case of independent data by using some probabilistic inequalities on moments of sums of i.i.d. variables. To deal with dependent data it is necessary to have some "equivalent" tool. The following result can be seen as an extension, in a very special case, of the inequalities given in theorems 2.2.4, 2.2.5, and 2.2.6. This result concerns real kernel functions K satisfying the following properties

$$\exists \bar{K},\ 0<\bar{K}<\infty,\ \forall x\in\mathbb{R},\ \|K(x)\| \leq \bar{K}; \qquad \text{(K.7)}$$

K has a absolutely integrable Fourier transform. (K.8)

K is compactly supported; (K.9)

K is symmetric

and (K.10)

$\int K(x)dx=1$.

Let us introduce the simpler notation

$$K_h(a,b) = 1/h\ K((a-b)/h).$$

Theorem 6.2.1. (Hart and Vieu, 1988). Let K be a kernel function satisfying the conditions (K.7)-(K.10), and assume that the sequence $(X_i)_N$

is α-mixing. Let $j(1),\ldots,j(p)$ be p distinct positive integers, and define

$$g(X_{j(1)},\ldots,X_{j(p)})=\prod_{\substack{r,s=1\\r\neq s}}^{q} K_h(X_{j(r)}-X_{j(s)})^{\beta_{r,s}} \prod_{i=1}^{p} g_{j(i)}(X_{j(i)}),$$

where the $g_{j(i)}$ are real valued functions such that $|g_{j(i)}(.)|\leq M_i<\infty$, the $\beta_{r,s}$ are nonnegative integers, and $q\leq p$. Let $(A_i)_{i=1}^{i=v}$ be a partition of $\{j(1),\ldots,j(p)\}$. Then, there exists a finite real positive constant G such that we have

$$|\int g dP_{(X_{j(1)},\ldots,X_{j(p)})}-dP_{(X_t,t\in A_1)},\ldots,dP_{(X_t,t\in A_v)}|\leq G\bar{\alpha}_d \prod_{i=1}^{p} M_i h^{-\beta},$$

where

$$\bar{\alpha}_d = \sup_{j\geq d} \alpha(d),$$

$$d = \inf \{d(A_i,A_j),\ i,j=1,\ldots v,\ i\neq j\},$$

$$d(A_i,A_j) = \inf \{|u-u'|,\ u\in A_i,\ u'\in A_j\},$$

and

$$\beta = \sum_{\substack{r,s=1\\r\neq s}}^{q} \beta_{r,s}.$$

This theorem was proven in Hart and Vieu (1988) in a sketched form. That is one of the reasons why we give it in great details here. The other reason, is that this result will be used later for statements of results in chapter 6. The authors wish to thank Jeffrey Hart who played an important role in the outlining of this proof.

<u>Proof of theorem 6.2.1.</u> In order to get simpler expressions the proof is only given in the case when $A_i=\{j(i)\}$, $\forall i=1,\ldots,p$. It would perform axactly over the same steps in case of more complicated partitions. Let us first look at the following quantity:

$$I = \prod_{\substack{r,s=1\\r\neq s}}^{q} K_h(X_{j(r)}-X_{j(s)})^{\beta_{r,s}}.$$

We have, by (K.11),

$$I = \prod_{\substack{r,s=1\\ r\neq s}}^{q} \left[\frac{1}{2\pi} \int_{-\infty}^{+\infty} \xi(hu) \exp\{-iu(X_{j(r)}-X_{j(s)})\}\, du\right]^{\beta_{r,s}},$$

$$=2\pi^{-\beta} \prod_{\substack{r,s=1\\ r\neq s}}^{q} \int\ldots\int \prod_{j=1}^{\beta_{r,s}} \xi(hu_j^{r,s})\exp\{-i(X_{j(r)}-X_{j(s)}) \sum_{j=1}^{\beta_{r,s}} u_j^{r,s}\} \prod_{j=1}^{\beta_{r,s}} du_j^{r,s},$$

$$=2\pi^{-\beta}\int\ldots\int \prod_{\substack{r,s=1\\ r\neq s}}^{q} \prod_{j=1}^{\beta_{r,s}} \xi(hu_j^{r,s}) \prod_{\substack{r,s=1\\ r\neq s}}^{q} \exp(i(X_{j(s)}-X_{j(r)}) \sum_{j=1}^{\beta_{r,s}} u_j^{r,s}) \prod_{\substack{r,s=1\\ r\neq s}}^{q} \prod_{j=1}^{\beta_{r,s}} du_j^{r,s}$$

Let us define

$$v(r,s) = \sum_{j=1}^{\beta_{r,s}} u_j^{r,s},$$

$$V_r = \sum_{\substack{s=1\\ r\neq s}}^{q} v(r,s)$$

and

$$V'_s = \sum_{\substack{r=1\\ r\neq s}}^{q} v(r,s).$$

We have

$$\prod_{\substack{r,s=1\\ r\neq 1}}^{q} \exp(-i(X_{j(r)}-X_{j(s)})v(r,s))$$

$$= \exp(-i\sum_{r\neq s}\sum (X_{j(r)}-X_{j(s)})v(r,s))$$

$$= \exp(-i(\sum_{r=1}^{q} X_{j(r)}V_r - \sum_{s=1}^{q} X_{j(s)}V'_s))$$

$$= \exp(-i \sum_{r=1}^{q} X_{j(r)}(V_r - V'_r)).$$

We therefore get the following expression for I:

$$I = 2\pi^{-\beta} \int \ldots \int \prod_{\substack{r,s=1 \\ r \neq s}}^{q} \prod_{j=1}^{\beta_{r,s}} \xi(hu_j^{r,s}) \exp(-i \sum_{r=1}^{q} X_{j(r)}(V_r - V'_r)) \prod_{\substack{r,s=1 \\ r \neq s}}^{q} \prod_{j=1}^{\beta_{r,s}} du_j^{r,s}.$$

Let us look now at the quantity:

$$II = \left| \int g \, dP_{X_{j(1)},\ldots,X_{j(p)}} - dP_{X_{j(1)}} \ldots dP_{X_{j(p)}} \right|,$$

and let us define

$$dQ = dP_{X_{j(1)},\ldots,X_{j(p)}} - dP_{X_{j(1)}} \ldots dP_{X_{j(p)}}.$$

We have

$$II = (\frac{1}{2\pi})^{\beta} \left| \int \ldots \int \prod_{\substack{r,s=1 \\ r \neq s}}^{q} \prod_{j=1}^{\beta_{r,s}} \xi(hu_j^{r,s}) \left[\int \exp(-i \sum_{r=1}^{q} X_{j(r)}(V_r - V'_r)) \prod_{s=1}^{p} g_{j(s)}(X_{j(s)}) dQ \right] \times \prod_{\substack{r,s=1 \\ r \neq s}}^{q} \prod_{j=1}^{\beta_{r,s}} du_j^{r,s} \right|.$$

The integrand of the inner integral may be expressed as $\prod_{s=1}^{p} h_s(X_{j(s)})$, where

$$h_s(t) = \exp(-it(V_s - V'_s)) g_{j(s)}(t) \text{ for } 1 \leq s \leq q$$

and

$$h_s(t) = g_{j(s)}(t) \text{ for } s > q.$$

The inner integral is

$$III = \int \prod_{s=1}^{p} h_s(X_{j(s)})\, dQ.$$

Without loss of generality we can assume that $j(1)<j(2)< \ldots <j(p)$. Let r' be the index such that

$$j(r'+1) - j(r') = d.$$

The above integral is then

$$III=E[\prod_{s=1}^{p} h_s(X_{j(s)})]-E[\prod_{s=1}^{r'} h_s(X_{j(s)})]E[\prod_{s=r'+1}^{p} h_s(X_{j(s)})]$$

$$+ E[\prod_{s=1}^{r'} h_s(X_{j(s)})]E[\prod_{s=r'+1}^{p} h_s(X_{j(s)})] \qquad (6.2.1)$$

$$- \prod_{s=1}^{p} E[h_s(X_{j(s)})].$$

From theorem 2.2.5, the term on the first line of (6.2.1) is bounded in absolute value by

$$4 \prod_{s=1}^{p} M_s \alpha(d). \qquad (6.2.2)$$

Define now

$$Y_1 = \prod_{s=1}^{r'} h_s(X_{j(s)}),\ Y_2 = \prod_{s=r'+1}^{p} h_s(X_{j(s)}),$$

$$Y_1' = \prod_{s=1}^{r''} h_s(X_{j(s)}),\ \tilde{Y}_1 = \prod_{s=r''+1}^{r'} h_s(X_{j(s)}),\ (r''<r'),$$

$$Y_2' = \prod_{s=r'+1}^{\tilde{r}} h_s(X_{j(s)}) \text{ and } \tilde{Y}_2 = \prod_{s=\tilde{r}+1}^{p} h_s(X_{j(s)}),\ (r'+1\leq\tilde{r}<p).$$

Now, the term in the second line of (6.2.1) is

$$[E(Y_1) - E(Y_1')E(\tilde{Y}_1)]E(Y_2) + E(Y_1')E(\tilde{Y}_1)[E(Y_2) - E(Y_2')E(\tilde{Y}_2)]$$
$$+ E(Y_1')E(\tilde{Y}_1)E(Y_2')E(\tilde{Y}_2).$$

Using again theorem 2.2.5, we get

$$|E(Y_1) - E(Y_1')E(\tilde{Y}_1)|\,|E(Y_2)| \leq 4 \prod_{s=1}^{p} M_s\alpha_{d*},\ (d*\geq d),$$

and

$$|E(Y_1')E(\tilde{Y}_1)||E(Y_2) - E(Y_2')E(\tilde{Y}_2)| \le 4 \prod_{s=1}^{p} M_s\alpha_{d'}, \ (d'\ge d).$$

These last two inequalities, together with (6.2.2) lead to

$$III \le 4 \prod_{s=1}^{p} M_s \, [\alpha_d+\alpha_{d*}+\alpha_{d'}]$$

$$+ \left| E(Y_1')E(\tilde{Y}_1)E(Y_2')E(\tilde{Y}_2) - \prod_{s=1}^{p} E[h_s(X_{j(s)})] \right|.$$

It is now clear that the above procedure can be iterated to obtain

$$III \le C \prod_{s=1}^{p} M_s \sup_{j\ge k} \alpha_j. \tag{6.2.3}$$

(Let us remark that this bound for the integral III could be obtained directly from lemma 4.2 in Castellana and Leadbetter (1986).) Note that the constant C in the last inequality does not depend on the variables $u_j^{r,s}$. Hence, we get from (6.2.3),

$$II \le C \prod_{s=1}^{p} M_s \sup_{j\ge d} \alpha_j \int\cdots\int \prod_{\substack{r,s=1\\ r\ne s}}^{q} \prod_{j=1}^{\beta_{r,s}} |\xi(hu_j^{r,s})| \prod_{\substack{r,s=1\\ r\ne s}}^{q} \prod_{j=1}^{\beta_{r,s}} du_j^{r,s},$$

and then

$$II = C \prod_{s=1}^{p} M_s \sup_{j\ge d} \alpha_j \, [\int_{-\infty}^{+\infty} |\xi(hu)|du]^{\beta}.$$

By using (K.11) we finally get

$$II \le Ch^{-\beta}[\int_{-\infty}^{+\infty} |\xi(u)|du]^{\beta} \prod_{s=1}^{p} M_s \sup_{j\ge k} \alpha_j,$$

which completes the proof of theorem 6.2.1.

3. Cross-validation for kernel density estimates.

3.1. Introduction.

We consider the same problem as in chapter IV but in an univariate

setting, namely the estimation of the density function f of a **real** random variable X from a sample $X_1,\dots,X_n$ of realisations of X. The sequence $(X_i)_N$ will be assumed to satisfy the α-mixing condition (see definition 2.2.3). Our

estimate is the same as defined in (4.2.1),

$$f_n(x) = 1/(nh) \sum_{i=1}^{n} K((x-X_i)/h),$$

K being a real kernel function and h a positive real smoothing parameter depending on n. We use as a measure of accuracy for this estimate the following integrated square error

$$ISE = ISE(h) = \int (f_n(x) - f(x))^2 w(x)dx, \qquad (6.3.1)$$

where w is some nonnegative weight function. Our aim is to select a data-driven bandwidth $\hat{h}$ which is asymptotically optimal in the sense of the definition 6.1.1. A reasonable way to do that is to consider the cross-validation criterion

$$CV_{l_n}(h) = \int f_n(x)^2 w(x)dx - 2n^{-1} \sum_{i=1}^{n} f_n^i(X_i)w(X_i). \qquad (6.3.2a)$$

The quantity f_n^i is defined by

$$f_n^i(x) = (h\bar{\gamma}_i)^{-1} \sum_{j=1}^{n} K((x-X_j)/h)\gamma(i-j), \qquad (6.3.2b)$$

where γ is a function satisfying

$$\gamma(0) = 0,$$
$$\gamma(x) = 1, \text{ if } x>l_n, \qquad (6.3.2c)$$
$$0\le\gamma(x)\le 1, \text{ if } x\le l_n,$$

$(l_n)_N$ being a sequence of positive integers, called the leave-out-sequence, and $\bar{\gamma}_i$ is such that

$$\bar{\gamma}_i = \sum_j \gamma(i-j). \qquad (6.3.2d)$$

This criterion is a smooth extension of the leave-out-procedure which consists in taking as particular function γ the following one

$$\gamma = 1 - I_{[-l_n, +l_n]}. \tag{6.3.3}$$

This particular case gives explanations about why we call (l_n) the leave-out- sequence. The role of the function γ is quite similar to the role of the kernel function K. While K was used to classify the data following their closeness in space, γ classifies these data following their closeness in time. The leave-out sequence $(l_n)_N$ determines roughly the distance (in time) after which two data can be reasonably treated as if they were independent. Note that the usual cross- validation criterion, introduced for independent data by Bowman (1984) and Rudemo (1982), corresponds to the case when when we just leave-out one data (i.e., $l_n=0$ for any n). In section 3.2 below we give asymptotic optimality of the bandwidth which minimises this criterion under the α-mixing condition on the data $(X_i)_N$. In corollary 6.3.1, it will be seen that when $l_n=0$ we still have asymptotic optimality; this is an extension of results of Hall (1983) and Marron (1987) who gave asymptotic optimality for $l_n=0$ when the data are independent.

3.2. Asymptotic optimality of the cross-validation procedure.

In order to get an asymptotic optimality property we make the following assumptions. The nonparametric model is defined by the following restrictions.

f has k continuous derivatives $(k>0)$; (A.5)

$\max(f(x), f(-x)) \text{ -----> } 0$ as $x \text{ -----> } \infty$; (A.6)

$\exists M_1,\ 0<M_1<\infty,\ \forall x \in \mathbb{R},\ f(x) \le M_1$; (A.7)

For any j, (X_j, X_{j+1}) has a density f_j with respect to Lebesgue measure. (A.8)

The conditions on the kernel function K are those described in section 2 of this chapter and in addition it is assumed that

$0<|\int t^k K(t)dt|<\infty$,

and (K.11)

$\int |t|^\gamma K(t)dt=0,\ \gamma=1,\ldots,k-1.$

The weight function w is such that

w is bounded and S = support(w) is compact, (W.1)

The selected bandwidth is

$$\hat{h}_{l_n} = \arg\min_{h \in H_n} CV_{l_n}(h), \qquad (6.3.4)$$

where

H_n is a finite subset of $[An^{-a}, Bn^{-b}]$,

for some $0 < b \leq 1/(2k+1) \leq a < 2/(1+4k)$, (H.11)

A and B being finite real positive constants.

The first result that we present is asymptotic optimality of the bandwidth $\hat{h}_{l_n}$ when the leave-out sequence $(l_n)_N$ is large enough, i.e. when this sequence satisfies together with the mixing coefficients α_n

$$l_n = n^{\tau_1} \text{ for some } 0 < \tau_1 < (2-a(1+4k))/2, \qquad (L.1)$$

and

$$\sup_{j \geq l_n} \alpha_j = o(n^{-\tau_2}) \text{ for } \tau_2 = U+V+(2a+4ka)(2+U/V), \qquad (L.2)$$

where

$U=1+2a+2ka-b$ and $V=2-a(1+4k)-2\tau_1$.

Theorem 6.3.1. Optimality of the leave-out procedure over a discrete set of smoothing parameter. (Hart and Vieu, 1988). Assume that the sequence $(X_i)_N$ is α-mixing and that conditions (H.11), (W.1), (L.1), (L.2), (K.7)-(K.11) and (A.5)-(A.8) hold. Then, the cross-validated bandwidth defined by (6.3.2)-(6.3.4) is asymptotically optimal with respect to integrated squarre error, in the sense that

$$ISE(\hat{h}(l_n))/(\inf_{h \in H_n} ISE(h)) \xrightarrow{a.s.} 1.$$

Proof of Theorem 6.3.1. The asymptotic optimality property follow from

$$\sup_{h \in H_n} |CT_{l_n}(h)|/ISE(h) \xrightarrow{a.s.} 0, \qquad (6.3.5)$$

where

$$CT_{l_n}(h) = n^{-1} \sum_{i=1}^{n} f_n^i(X_i)w(X_i) - \int f(x)f_n(x)w(x)dx \qquad (6.3.6)$$

$$- n^{-1}\sum_{j=1}^{n} f(X_j)w(X_j) + \int f^2(x)w(x)dx.$$

Therefore, it remains only to check that

$$\sup_{h\in H_n} |CT_{1_n}(h)|/ISE(h) \xrightarrow{a.s.} 0, \quad (6.3.7)$$

The error ISE is basically the sum of two positive components (see e.g. Rosenblatt (1971a)), i.e, a variance component and a squared bias component. The squared bias component

$$b(h) = \int (Ef_n(x)-f(x))^2 w(x)dx,$$

does not depend on the multidimensionnal distribution of $X_1,\ldots,X_n$, and is therefore the same as for independent samples. Following the techniques previously described by Parzen (1962) in the case $\gamma=2$, Collomb (1976) showed by Taylor expansion of f, that under the conditions (A.5), (W.1) and (K.11), b(h) is asymptotically of the same order as h^{2k}. Therefore, there exists a finite positive constant C_b such that for n large enough we have for any h in $[An^{-a}, Bn^{-b}]$,

$$ISE(h) \geq C_b h^{2k}. \quad (6.3.8)$$

Therefore, to get (6.3.7) it is enough to show

$$\sup_{h\in H_n} |CT_{1_n}(h)| h^{-2k} \xrightarrow{a.s.} 0, \quad (6.3.9)$$

Let us now write

$$CT_{1_n}(h) = n^{-1} \sum\sum_{|i-j|>1_n} (\bar{\gamma}_i)^{-1} \Gamma(i,j),$$

where

$$\Gamma(i,j) = K_h(X_i-X_j)w(X_j) - \int f(x)K_h(x-X_i)w(x)dx$$

$$- f(X_j)w(X_j) + \int f^2(x)w(x)dx.$$

Now, define

$$\Gamma^*(j) = \int K_h(u-X_j)w(X_j)f(u)du - \iint K_h(x-u)f(x)w(x)f(u)dudx$$
$$- f(X_j)w(X_j) + \int f^2(x)w(x)dx,$$

$$\Psi(i,j) = \Gamma(i,j) - \Gamma^*(j),$$

and

$$T(h) = \sum\sum_{|i-j|>l_n} \Psi(i,j).$$

Noting that $\sup_i \bar{\gamma}_i = O(n)$, and because of lemma 6.3.3 below, all we have to prove is that

$$\sup_{h\in H_n} n^{-2}h^{-2k} |T(h)| \xrightarrow{a.s.} 0. \qquad (6.3.10)$$

<u>Lemma 6.3.1.</u> (Hart and Vieu, 1988). Under the conditions of theorem 6.3.1, we have

$$\sup_{h\in H_n} n^{-1}h^{-2k} \left|\sum_{j=1}^{n} \Gamma^*(j)\right| \xrightarrow{a.s.} 0.$$

It remains now to prove (6.3.10). For this decompose the term T(h) in the following way:

$$T(h) = T^+(h) + T^-(h),$$

where

$$T^+(h) = \sum\sum_{i+l_n<j\le n} \Psi(i,j),$$

and

$$T^-(h) = \sum\sum_{1\le j<i-l_n} \Psi(i,j).$$

What we have to check is that we have

$$\sup_{h\in H_n} n^{-2}h^{-2k} |T^+(h)| \xrightarrow{a.s.} 0, \qquad (6.3.11)$$

and

$$\sup_{h\in H_n} n^{-2}h^{-2k} |T^-(h)| \xrightarrow{a.s.} 0. \qquad (6.3.12)$$

We only give the proof of (6.3.11), the one for (6.3.12) can be carried out similarly. Let us define $\Psi(i,j)=0$ when (i,j) is not in the set $\{(i,j),$

$i+l_n<j\leq n\}$, and write

$$T^+(h) = \sum_{s=0}^{1} \sum_{t=0}^{1} \sum_{j_1=1}^{l_n} \sum_{j_2=1}^{l_n} \sum_{q=1}^{n_1} \sum_{m=1}^{q} \Psi(m',q')$$

where for s,t, j_1 and j_2 fixed we use the simpler notations

$$m' = j_2+2(m-1)l_n+sl_n, \quad q' = j_1+(2q-1)l_n+tl_n,$$

and where n_1 is the greater integer less than or equal to $n/2l_n$. Let us now compute for some integer p the moments of order 2p of T^+. By Minkowski's inequality we have

$$E(T^+(h)^{2p})\leq(2l_n)^{4p} \sup_{j_1,j_2,s,t} E((\sum_{q=1}^{n_1} \sum_{m=1}^{q} \Psi(m',q'))^{2p}). \quad (6.3.13)$$

<u>Remark 6.3.1.</u> As soon as m_1 differs from m_2 we have $|m_1'-m_2'|>l_n$. Similarly, as soon as q_1 differs from q_2 we have $|q_1'-q_2'|>l_n$. And finally, as soon as m differs from q and from q+1 we have $|m'-q'|>l_n$.

We have to look at the following moments (which exist by (A.5)-(A.7))

$$E((\sum_{q=1}^{n_1} \sum_{m=1}^{q} \Psi(m',q'))^{2p})= \sum_{q_1=1}^{n_1} \sum_{m_1=q_1}^{q_1} \dots \sum_{q_{2p}=1}^{n_1} \sum_{m_{2p}=1}^{q_{2p}} E \prod_{i=1}^{2p} \Psi(m_i',q_i').$$

Let us now define

$$I = \{J=(m_1,q_1,\dots,m_{2p},q_{2p}),\ \forall i=1,\dots,2p,\ 1\leq m_i\leq q_i\leq n_1\},$$

and consider the subset I_1 of I composed of the 4p-uplets $J=(m_1,q_1,\dots,m_{2p},q_{2p})$ for which at least one index among $(m_1,q_1,\dots,m_{2p},q_{2p})$ differs from all other indices by at least 2. So we have to compute

$$E((\sum_{q=1}^{n_1} \sum_{m=1}^{q} \Psi(m',q'))^{2p}) = \sum_{J\in I_1} EZ(J) + \sum_{J\in I-I_1} EZ(J). \quad (6.3.14)$$

where for $J=(m_1,q_1,\dots,m_{2p},q_{2p})$, $Z(J)$ is defined by

$$Z(J) = \prod_{i=1}^{2p} \Psi(m_i',q_i').$$

Let first consider $J=(m_1,q_1,\dots,m_{2p},q_{2p})$ in I_1, and assume that m_{i_0} is the index which differs from each other by at least 2 (the proof would be similar if this index was some q_{j_0}). The idea is to expand $Z(J)$ as a finite sum of terms of form (4.3) and by applying theorem 6.2.1 to each of these terms by taking (using the notations introduced in this theorem):

$$v = 3,\ \beta = 2p,$$

$$A_1 = \{i\in\{m'_1,q'_1,\dots,m'_{2p},q'_{2p}\},\ i < m'_{i_0}\}$$

$$A_2 = \{m'_{i_0}\},$$

$$A_3 = \{i\in\{m'_1,q'_1,\dots,m'_{2p},q'_{2p}\},\ i > m'_{i_0}\},$$

noting that by remark 6.3.1 we have

$$d \geq l_n.$$

Finally, theorem 6.2.1 gives for some finite positive real constant C_1

$$|EZ(J)|\leq C_1 h^{-2p}\sup_{j\geq l_n}\alpha_j+\left|\iiint Z(J)dP_{X_{m'_{i_0}}}dP_{(X_t,t\in A_1)}dP_{(X_t,t\in A_3)}\right|,$$

and it suffices to note that, by definition of $\Psi(i,j)$, the second member in right hand side of the above inequality is 0 to conclude that

$$|EZ(J)| \leq C_1 h^{-2p}\sup_{j\geq l_n}\alpha_j \text{ for any } J\in I_1.$$

Noting that

$$\# I_1 = O(n_1^{4p}) \text{ and } \#(I-I_1) = O(n_1^{2p}), \text{ we finally get}$$

$$\left|\sum_{J\in I_1} EZ(J)\right| \leq O(n_1^{4p}h^{-2p}\sup_{j\geq l_n}\alpha_j). \tag{6.3.15}$$

Let us now consider indices J in the set $I-I_1$. For this define the following partition of $I-I_1$:

$$I-I_1 = \bigcup_{\gamma=1}^{4k} I^\gamma,$$

where

$$I^\gamma=\{J=(m_1,q_1,\dots,m_{2p},q_{2p})\in I-I_1,\ \#\{m_1,q_1,\dots,m_{2p},q_{2p}\}=\gamma\}.$$

Let $J=(m_1,q_1,\ldots,m_{2p},q_{2p})$ be in the set I^γ, and denote by $m_1,\ldots,m_{\gamma_1}$, $m_{\gamma_1+1},\ldots,m_{\gamma_2},q_1,\ldots,q_{\gamma_1},q_{\gamma_1+1},\ldots,q_{\gamma_3}$ (where $\gamma_2+\gamma_3=\gamma$) the γ distincts elements of $\{m_1,q_1,\ldots,m_{2p},q_{2p}\}$ classified in such a way that

for $r=1,\ldots,\gamma_1$, $m_r=q_r+1$,
for $r=\gamma_1+1,\ldots,\gamma_2$, $\forall j$, $m_r\neq 1+q_j$,
for $r=\gamma_1+1,\ldots,\gamma_3$, $\forall j$, $q_r\neq 1+m_j$.

Apply now theorem 6.2.1 with, using the notations of this theorem,

$$v = \gamma-\gamma_1,\ \beta = 2p,$$
$$A_r = \{m_r',q_r'\},\ \text{for } r=1,\ldots,\gamma_1,$$
$$A_r = \{m_r'\},\ \text{for } r=\gamma_1+1,\ldots,\gamma_2,$$
$$A_r = \{q'_{r-\gamma_2+\gamma_1}\}\ \text{for } r=\gamma_2+1,\ldots,\gamma-\gamma_1,$$

and note that remark 6.3.1 implies that $d\geq l_n^*$, to get for some finite positive constant C_2

$$|EZ(J)| \leq C_2 h^{-2p} \sup_{j\geq l_n} \alpha_j + \left|\int Z(J) \prod_{r=1}^{v} dP_{(X_t,\, t\in A_r)}\right|.$$

By using (A.8) and then by classical integration by substitution we get

$$\left|\int Z(J) \prod_{r=1}^{v} dP_{(X_t,\, t\in A_r)}\right| = O(h^{-2k+\gamma/2}).$$

Note that this last result was given by Marron and Hardle (1986, formula 3.4) but with $\prod_{i=1}^{\gamma} dP_{X_{t_i}}$ in place of $\prod_{r=1}^{v} dP_{(X_t,\, t\in A_r)}$ (see also Mielniczuk, Sarda and Vieu (1988)). The last two results lead to

$$|EZ(J)| = O(h^{-2p} \sup_{j\geq l_n} \alpha_j) + O(h^{-2p+\gamma/2}),\ \text{for any } J\in I^\gamma.$$

Noting that we have

$$\# I^\gamma = O(n_1^\gamma),$$

and that we have also, because $I^\gamma \subset I-I_1$,

$$\#I^{\gamma} = O(n_1^{2p}),$$

we have for some finite real positive constant C_3

$$\Big|\sum_{J\in I-I_1} EZ(J)\Big| \le C_3 h^{-2p}\Big(n_1^{2p}\sup_{j\le l_n}\alpha_j + \sum_{\gamma=1}^{2p} n_1^{\gamma} h^{\gamma/2} + \sum_{\gamma=2p+1}^{4p} n_1^{2p} h^{\gamma/2}\Big).$$

Noting that in the first (resp. in the second) sum on right hand side of this inequality the biggest term is for $\gamma=2p$ (resp. $\gamma=2p+1$), we get

$$\Big|\sum_{J\in I-I_1} EZ(J)\Big| = O(n_1^{2p}h^{-p}) + O(n_1^{2p}h^{-2p}\sup_{j\le l_n}\alpha_j). \qquad (6.3.16)$$

It follows from (6.3.13)-(6.3.16) and the fact that n_1 is of the same order as n/l_n, that

$$E((T^{+}(h))^{2p}) = O(n^{4p}h^{-2p}\sup_{j\le l_n}\alpha_j) + O(n^{2p}(l_n)^{2p}h^{-p}). \quad (6.3.17)$$

Now, using Chebyshev's inequality we have

$$P(\sup_{h\in H_n} n^{-2}h^{-2k}|T(h)|>\epsilon) \le$$

$$\#H_n'\epsilon^{-2p}\sup_{h\in H_n} E(n^{-2}h^{-2k}T(h))^{2p}. \qquad (6.3.18)$$

Finally, by using (L.1), (L.2), (H.11), (6.3.6), (6.3.17) and (6.3.18) and by taking p such that

$$1 + U/V < p \le 2 + U/V,$$

we get (6.3.11) which completes the proof of theorem 6.3.1.

We will now give extensions of theorem 6.3.1 to the case when H_n is a continuous set of smoothing parameter. Even if in practical situations we always do some discretization over the set of smoothing parameter, it is interesting to get theoretical results over all the continuous set not to have to worry with the problem of the cardinality of H_n. In order to get asymptotic optimality over the continuous set of smoothing parameter

$$H_n=[An^{-a},Bn^{-b}],\ 0<b\le 1/(2k+1)\le a<2/(1+4k), \qquad (H.12)$$

we need to assume the following additional assumption on the kernel function K

K is Lipschitz continuous, i.e.,

$$\exists C_K,\ 0<C_K<\infty,\ |K(x)-K(y)| \le C_K|x-y|; \tag{K.12}$$

Theorem 6.3.2. Extension to continuous set of smoothing parameters. (Hart and Vieu, 1988). Assume that the sequence $(X_i)_N$ is α-mixing and that conditions (H.12), (W.1), (L.1), (L.2), (K.7)-(K.12) and (A.5)-(A.8) hold. Then, the cross-validated bandwidth defined by (6.3.2)-(6.3.4) is asymptotically optimal with respect to integrated squarre error, in the sense that

$$ISE(\hat{h}(l_n))/(\inf_{h\in H_n} ISE(h)) \xrightarrow{a.s.} 1.$$

Proof of theorem 6.3.2. Let us denote by H'_n a finite subset of H_n composed of equally spaced elements, and such that

$$\#H'_n = n^{2a+2ka-b+\xi},\ \text{for some } 0<\xi<V.$$

We have the following result.

Lemma 6.3.2. (Hart and Vieu, 1988). Under the conditions of theorem 6.3.2 we have

$$\sup_{h\in H_n} |CT_{l_n}(h) - CT_{l_n}(h^*)|/ISE(h) \xrightarrow{a.s.} 0,$$

where, for h in H_n, h^* denotes the element of H'_n that is closest to h.

Now, as before, asymptotic optimality will be stated if we check (6.3.5). To show (6.3.5) for H_n it suffices to note that it is satisfied for H'_n (because of theorem 6.3.1) and to use lemma 6.3.2 above. This completes the proof of theorem 6.3.2.

The above theorems concern only the case when γ is the indicator function as defined in (6.3.3) and when the leave-out-sequence $(l_n)_N$ is large enough. Defining γ by (6.3.3) means that to construct the leave-out estimates f_n^i we suppress something like $2l_n$ data of the sample. Obviously, in situations when the sample size is quite small, this may reduce the value of our procedure. In the following result the function γ is not necessarly of the form (6.3.3) but only assumed to satisfy (6.3.2c). Also, the conditions on $(l_n)_N$ are now less restrictive, i.e.,

$$l_n \le n^{\tau_1} \text{ for some } 0<\tau_1<(2-a(1+4k))/2, \tag{L.3}$$

and

$$\sup_{j\geq n^{\tau_1}} \alpha_j = o(n^{-\tau_2}) \text{ for } \tau_2=U+V+(2a+4ka)(2+U/V). \qquad (L.4)$$

Theorem 6.3.3. Optimality of the "time smoothing" version of leave-out-procedure. Assume that the sequence $(X_i)_N$ is α-mixing and that conditions (H.12), (W.1), (L.3), (L.4), (K.7)-(K.12) and (A.5)-(A.8) hold. Then, the cross-validated bandwidth defined by (6.3.2) and (6.3.4) is asymptotically optimal with respect to integrated squarre error, in the sense that

$$ISE(\hat{h}(l_n))/(\inf_{h\in H_n} ISE(h)) \xrightarrow{a.s.} 1.$$

Proof of theorem 6.3.3. Let us denote by

$$l_n^* = n^{\tau_1},$$

and by γ^*, $\bar{\gamma}_i^*$ and $CT_{l_n^*}$ the quantities defined in (6.3.2d), (6.3.3) and (6.3.6) for l_n^*. As in theorem 6.3.1, it suffices to show (6.3.5). Because of theorem 6.3.2 all we have to prove is that

$$\sup_{h\in H_n} |CT_{l_n}(h)-CT_{l_n^*}(h)|/h^{2k} \xrightarrow{a.s.} 0. \qquad (6.3.19)$$

Write now

$$|CT_{l_n}(h) - CT_{l_n^*}(h)| \leq |R_1(h)| + |R_2(h)|$$

where

$$R_1(h) = \sum_{l_n<|i-j|<l_n^*}\sum (n\bar{\gamma}_i)^{-1}K_h(X_i-X_j)w(X_i)$$

and

$$R_2(h) = \sum_{|i-j|>l_n^*}\sum [(n\bar{\gamma}_i)^{-1}-(n\bar{\gamma}_i^*)^{-1}]K_h(X_i-X_j)w(X_i).$$

We can rewrite R_1 in the following form

$$R_1(h) = \sum_{|k|=l_n}^{|k|=l_n^*} \sum_{i=1}^{n} (n\bar{\gamma}_i)^{-1}K_h(X_i-X_{i-k})w(X_{i-k})$$

and, denoting by f_n^* the kernel estimate of f using as kernel function

$$K^* = |K| / \int |K(u)| du$$

we finally get (noting also that $\sup_{i=1,\dots,n} \bar{\gamma}_i$ is of the same order as n)

$$\sup_{h \in H_n} |R_1(h)| \leq 2n^{-1} \int |K(u)| du (l_n^* - l_n) \sup_{u \in S} f_n^*(u).$$

(Remember that S is the support of the weight function w.). Theorem 5.3.3 together with this last formula gives

$$\sup_{h \in H_n} |R_1(h)| = O(n^{-1} l_n^*).$$

By the same kind of arguments we get also

$$\sup_{h \in H_n} |R_2(h)| = O(n^{-1} l_n^*).$$

These two last results together with (H.12) and (L.3) lead to (6.3.19). This completes the proof of theorem 6.3.3.

In fact theorem 6.3.3 define a whole class of optimal data-driven smoothing parameter, indexed by the leave-out sequence $(l_n)_N$. (It is quite intuitive, as it is the case for the kernel function K, that the shape of the function γ will be very much less important than the value of parameter l_n.) There is two obvious questions that one may want to answer

-How choose l_n in practical situations?

-Is the usual version of cross-validation, i.e. $l_n=0$ for any n, still optimal for dependent data?

While we don't have theoretical answer to the first question (partial element of answer are discussed through simulated examples by Hart and Vieu, 1988), our conditions on $(l_n)_N$ allows us to treat the case of ordinary cross-validation (i.e. $l_n=0$, for any n) as indicated in the following result.

__Corollary 6.3.1.__ **Asymptotic optimality of the usual leave-one-out procedure.** (Hart and Vieu, 1988). Under the conditions of theorem 6.3.3, the "ordinary" version of cross-validation, i.e. the case when

$$l_n = 0 \text{ for any } n,$$

is asymptotically optimal with respect to integrated square error, according to the definition 6.1.1.

__Remark 6.3.2.__ This last result was given by Hall (1983) and Marron (1987) in the setting of independent data.

__Remark 6.3.3.__ We suspect that all the results of this section remain valid for other quadratic measure of errors, as for example averaged square error ASE,

$$ASE = ASE(h) = n^{-1} \sum_{i=1}^{n} (f(X_i) - f_n(X_i))^2 w(X_i),$$

or mean integrated square error MISE,

$$MISE = MISE(h) = E(ISE(h)).$$

In the setting of independent data, Marron and Härdle (1986) have proven that all these quadratic measures of errors were asymptotically equivalent, but we do not know a similar result when the data are dependent.

__3.3. About other smoothing parametric techniques.__

Until now, only the cross-validation procedure, studied above, has been investigated when the data are not independent. Several other smoothing parameter selection rules have already been introduced in the setting of independent data. We did not present the details of these procedures since the main purpose of this book concerns dependent data. However, we want to mention references about what exists in the i.i.d. case, with the wish that it will motivate further researches on the behaviour of these procedures in the case of dependent data.

The reader will find in Rice (1984) several other data-driven criterions that could be used in order to get asymptotically optimal (with respect to quadratic measure of error) bandwidth. Let us also note, that a local version of this cross-validation method have been introduced by Vieu (1988) and applied to density estimation by Mielniczuk et al. (1988) and Hall and Schucany (1988). This local cross-validation procedure allows to get location adaptive data-driven bandwidths which are asymptotically optimal with respect to some local version of quadratic errors.

If in place of looking for minimising quadratic error we investigate Kullback Leibler loss, Hall (1987 a and b) proposes another bandwidth selection technique and shows its asymptotic optimality. The reader will find elements of comparison between Hall's method and the cross-validation method above studied in Marron (1987).

A large survey of all these procedures is presented by Marron (1988).

4 Cross-validation for kernel hazard estimates.

4.1. Introduction.

We consider the same problem as in chapter V but in **an univariate setting**, namely the estimation of the hazard function g of a **real** random variable X from a sample $X_1,\dots,X_n$ of realisations of X. The sequence $(X_i)_N$ will be assumed to satisfy
the α-mixing condition (see definition 2.2.3). The estimates to be considered are the kernel hazard estimates introduced in chapter 5

$$g_n(x) = f_n(x)/(1-F_n(x)),\ \forall x\ F_n(x) < 1,$$

where

$$f_n(x) = 1/(nh) \sum_{i=1}^{n} K((x-X_i)/h),$$

$$F_n(x) = 1/n \sum_{i=1}^{n} I_{\{X_i \le x\}},$$

and where K is a real kernel function and h a real positive smoothing parameter. Our measure of accuracy is the following integrated square error

$$ISE = ISE(h) = \int (g_n(x) - g(x))^2 w(x)dx, \tag{6.4.1}$$

w being some positive weight function. In order to choose a data-driven bandwidth which asymptotically minimises this measure of error, we introduce the following cross-validation criterion

$$CV_{1_n}(h) = \int g_n^2(x)w(x)dx - 2n^{-1} \sum_{i=1}^{n} \frac{f_n^i(X_i)}{(1-F_n(X_i))^2} w(X_i), \tag{6.4.2}$$

The quantity f_n^i is defined by

$$f_n^i(x) = (h\bar{\gamma}^i)^{-1} \sum_{|j-i|>l_n}^{n} K((x-X_j)/h)\gamma(i-j), \tag{6.4.3}$$

where the function γ is such that,

$$\gamma(0) = 0,$$
$$\gamma(x) = 1, \text{ if } x > l_n, \tag{6.4.4}$$

$0 \leq \gamma(x) \leq 1$, if $x \leq l_n$,

where $(l_n)_N$ is a sequence of positive integers, called the leave-out-sequence, and where

$$\bar{\gamma}_i = \sum_{j=1}^{n} \gamma(i-j). \qquad (6.4.5)$$

The selected bandwidth is

$$\hat{h}_{l_n} = \arg\min_{h \in H_n} CV_{l_n}(h), \qquad (6.4.6)$$

where H_n satisfies the same condition as for density estimation (see section 3 of this chapter), namely it is such that for some constants A, B, a and b we have

$$H_n = [An^{-a}, Bn^{-b}],\ 0 < b \leq 1/(2k+1) \leq a < 2/(1+4k), \qquad (H.12)$$

recalling that k is the number of derivatives of f. The reader will find in Sarda and Vieu (1989,a and b) detailled motivations for the introduction of such a criterion.

4.2. Asymptotic optimality.

In order to get an asymptotic optimality property we need the same assumptions as in the setting of density estimation (see section 3.2 of this chapter), with an additional one in order to avoid the denominator of our estimate to vanish. This new assumption concerns the support S of the weight function w,

$$\exists \xi_1 > 0,\ \forall x \in S,\ F(x) \leq 1-\xi_1. \qquad (W.2)$$

Theorem 6.4.1. (Sarda and Vieu, 1989b). Assume that the sequence $(X_i)_N$ is α-mixing and that conditions (H.12), (W.1), (W.2), (L.3), (L.4), (K.7)-(K.12) and (A.5)-(A.8) hold. Then, the cross-validated bandwidth defined by (6.4.2)-(6.4.6) is asymptotically optimal with respect to integrated square error, in the sense that

$$ISE(\hat{h}(l_n))/(\inf_{h \in H_n} ISE(h)) \xrightarrow{a.s.} 1.$$

Proof of theorem 6.4.1. This proof is not given for two main reasons. The first one is that it follows roughly the same steps as those described in section 3 in density estimation (note that here also the crucial point is the utilisation of theorem 6.2.1). Just minor technical changes are needed to deal, at each step of the proofs, with the denominator of our hazard estimate, and it would be tedious to rewrite all the computations of section 3.2 just for these minor changes. The second reason is that all these computations are given in great details in Sarda and Vieu (1989a).

Remark 6.4.1. Condition (H.12) on the kernel function can be suppressed, and similarly to what happened in density estimation (compare theorem 6.3.1 and theorem 6.3.2) we could state our asymptotic optimality property without the condition (K.12). In counterpart, the condition would be changed in the stronger condition (H.11).

As in density estimation it remains two important questions to be answered, i.e.,

-How choose l_n?

-What happens in the usual case when $l_n=0$, $\forall n$?

and as in density estimation we are only able to answer the second one.

Remark 6.4.2. Conditions (L.3) and (L.4) allows us to take $l_n=0$ for any n. Therefore, the usual cross-validation procedure, i.e. the case when $l_n=0$ for any n, is still asymptotic optimal under the α-mixing condition. This procedure was introduced and shown asymptotically optimal in the i.i.d. case by Sarda and Vieu (1988a).

Remark 6.4.3. It would also be interesting to get similar result with other quadratic measures of errors, like for instance averaged square error ASE,

$$ASE = ASE\ (h) = n^{-1} \sum_{i=1}^{n} (g(X_i) - g_n(X_i))^2 w(X_i),$$

or mean integrated square error MISE,

$$MISE = MISE\ (h) = E(ISE(h)).$$

When the data are independent all these measures of accuracy have been seen to be equivalent by Marron and Hardle (1986) but we cannot offer an extension of their result to the dependent case.

5. Cross-validation for kernel regression estimates.

5.1. Introduction.

We concentrate here on the regression setting as defined in chapter III but in an **univariate setting**, namely the problem of the nonparametric estimation of a **real** regression function

$$r(.) = E(Y|X=.),$$

from a sample $(X_i,Y_i)_N$ of realisations of a pair of random variables (X,Y) valued in $\mathbb{R}^2$. Here we still use the kernel estimators introduced by Nadaraya (1964) and Watson (1964) and defined by

$$r_n(x) = \sum_{i=1}^{n} Y_i K((x-X_i)/h) / \sum_{i=1}^{n} K((x-X_i)/h),$$

where K is a kernel function defined on $\mathbb{R}$. The smoothing parameter is the bandwidth h (which depends on n), and its role is prominent in the behaviour of the regression estimate r_n (see the discussion in section 1 of the present chapter).

The most popular selection rule is the cross-validation method that was investigated by Hardle and Marron (1985) for i.i.d. variables and by HCrdle, Vieu and Hart (1989) for dependent data. The bandwidth is selected to minimise the following criterion:

$$CV_{l_n}(h) = n^{-1} \sum_{i=1}^{n} (Y_i - r_n^i(X_i))^2 w(X_i), \tag{6.5.1}$$

where w is some nonnegative weight function, and where

$$r_n^i(x)=(h\bar{\gamma}^i f_n^i(X_i))^{-1} \sum_{|j-i|>l_n}^{n} Y_j K((x-X_j)/h)\gamma(i-j), \tag{6.5.2a}$$

$$f_n^i(x)=(h\bar{\gamma}^i)^{-1} \sum_{|j-i|>l_n}^{n} K((x-X_j)/h)\gamma(i-j), \tag{6.5.2b}$$

where the function γ is such that,

$$\begin{aligned} &\gamma(0) = 0, \\ &\gamma(x) = 1, \text{ if } x>l_n, \\ &0"\gamma(x)\leq 1, \text{ if } x\leq l_n, \end{aligned} \tag{6.5.3}$$

where $(l_n)_N$ is a sequence of positive integers, called the leave-out-sequence, and where

$$\bar{\gamma}_i = \sum_{j=1}^{n} \gamma(i-j). \tag{6.5.4}$$

This funcion γ is introduced in such a way that data which are close in time to (X_i,Y_i) are given less weight than those which are farther away in time. This represents a compromise between the ordinary method of cross-validation introduced for independent data (see e.g. Hardle and Marron (1985)), i.e. the case when l_n=0 for any N, and the natural extension of this usual cross-validation rule that could be though by taking the particular weight function

$$\gamma(x) = I_{[-l_n, +l_n]}.$$

In fact, the role of γ is quite similar to that of the kernel function K. The function K classifies the data according to their closeness in space, while γ classifies them according to their closeness in time.

The bandwidth is selected to be the minimiser over the set H_n defined in (H.12) of the score function CV defined in (6.5.1), i.e,

$$\hat{h}_{l_n} = \arg\min_{h \in H_n} CV_{l_n}(h).$$

Let us denote by f the marginal density of X. In addition to the conditions needed to deal with cross-validation for density estimation (see section 3 of this chapter), we need to assume that

r has k continuous derivatives, (A.9)

and that,

$Y_i < C_4 < \infty$, for any $i = 1, \ldots, n$. (A.10)

Also to avoid the random denominator of our estimate to vanish we have to assume that, recalling that S is the compcat support of the weight function w,

f is bounded away from 0 on the interior of S. (W.3)

In the following theorem, we state asymptotic optimality (according the sense of definiton 6.1.1), of this method when the variables (X_i, Y_i) are α-mixing. This optimality result is stated by choosing as a measure of accuracy the following averaged square error,

$$ASE = ASE(h) = n^{-1} \sum_{i=1}^{n} (r(X_i) - r_n(X_i))^2 w(X_i).$$

Theorem 6.5.1. (Hardle, Vieu and Hart, 1989). Assume that the sequence $(X_i, Y_i)_N$ is **α-mixing** and that conditions (H.12), (W.1), (W.3), (L.3), (L.4), (K.7)-(K.12) and (A.5)-(A.10) hold. Then, the cross-validated bandwidth defined by (6.5.2)-(6.5.6) is asymptotically optimal with respect to averaged square error, in the sense that

$$ASE(\hat{h}(l_n)) / (\inf_{h \in H_n} ASE(h)) \xrightarrow{a.s.} 1.$$

Remark 6.5.1. Condition (K.12) on the kernel function can be suppressed, and similarly to what happened in density estimation (compare theorem 6.3.1 and theorem 6.3.2) we could state our asymptotic optimality property without the condition (K.12). In counterpart, the condition would be changed in the stronger condition (H.11).

As in density estimation it remains two important questions to be answered, i.e.,

-How choose l_n?

-What happens in the usual case when $l_n=0$, $\forall n$?

and as in density estimation we are only able to answer the second one.

Remark 6.5.2. Conditions (L.3) and (L.4) allows us to take $l_n=0$ for any n. Therefore, the usual cross-validation procedure, i.e. the case when $l_n=0$ for any n, is still asymptotic optimal under the α-mixing condition. This procedure was already shown to be asymptotically optimal in the i.i.d. case by HCrdle and Marron (1985).

Remark 6.5.3. It would also be interesting to get similar result with other quadratic measures of errors, like for instance integrated square error ASE,

$$ISE = ISE(h) = (r(x) - r_n(x))^2 w(x)dx,$$

or mean integrated square error MISE,

$$MISE = MISE\ (h) = E(ISE(h)).$$

When the data are independent all these measures of accuracy have been seen to be equivalent by Marron and Hardle (1986) but we cannot offer an extension of their result to the dependent case.

BIBLIOGRAPHY.

Abou-Jaoude, S. (1976a) Conditions nAcessaires et suffisantes de convergence L_1 en probabilité de l'histogramme pour une densité. Annales de l'Institut Henri Poincaré, 12, pp. 213-231.

Abou-Jaoude, S. (1976b) Sur une condition nécessaire et suffisante de L_1-convergence presque complète de l'estimateur de la partition fixe pour une densité. C.R. Acad. Sc. Paris, 283, pp. 1107-1110.

Azuma, K. (1967) Weighted sums of certain dependent random variables. Tohoku Math. J. 19, pp.357-367.

Bartlett, M.S. (1963) Statistical estimation of Density function. Sankhya A, 25, pp. 245-254.

Beck, A. (1963) On the strong law of large numbers. In: Ergodic Theory, F.B. Wright Ed., Academic Press, pp. 21-53.

Beck, J. (1979) The exponential rate of convergence of error for k-NN nonparametric regression and decision. Problems of Control and Information Theory, 8, pp. 303-312.

Berkes, I. and Philipp, W. (1977) An almost sure invariance principle for the empirical distribution function of mixing random variables. Zeitschrift f. W. u. v. G., 41, pp. 115-137.

Beyong Park and Marron J.S. (1988) Comparison of data-driven bandwidth selectors. Tech. Report 1759, Univ. North Carolina, Chapel Hill.

Bosq, D. (1973) Sur l'estimation de la densité d'un processus stationnaire et mélangeant. C.R. Acad. Sci. Paris, 277, pp. 535-538.

Bosq, D. (1975) Inégalité de Berstein pour les processus stationnaires et mélangeants Applications. C.R. Acad. Sci. Paris, 281, pp. 1095-1098.

Bosq, D. and Delecroix, M. (1985) Nonparametric prediction of a Hilbert space valued random variable. Stoc. Proc. and their Appl., 19, pp. 271-280.

Bowman, A. (1984) An alternative method of cross-validation for the smoothing of density estimates. Biometrika, 71, pp. 353-360.

Bradley, R.C. (1980a) On the φ-mixing condition for stationary random sequences. Duke Math. Journ., 47, 2, pp. 421-433.

Bradley, R.C. (1980b) On the strong mixing and weak Bernouilli conditions. Zeitschrift f. W. u. v. G., 51, pp. 49-54.

Bradley, R.C. (1980c) On the central limit question for dependent random variables. Jour. Appl. Proba., 17, pp. 94-101.

Bradley, R.C. (1980d) A note on strong mixing conditions. Annals of Probability., 8, pp. 636-638.

Bradley, R.C. (1981a) A sufficient condition for linear growth of variances in a stationary random sequence. Proc. of the Amer. Math. Soc., 83, 3, pp. 586-600.

Bradley, R.C. (1981b) Central limit theorems under weak dependence. Journ. of Multiv. Anal., 11, pp. 1-16.

Bradley, R.C. (1983) Asymptotic normality of some kernel-type estimators of probability density. Statistics & Probability Letters, 1, pp. 295-300.

Bradley, R.C. (1985) Some remarks on strong mixing conditions. Proc. of 7th Conf. on Proba. Theory, Brasov, Romania (1982), Ed. M. Losifescu, pp. 65-72.

Carbon, M. (1982) Sur l'estimation asymptotique d'une classe de parametres fonctionnels pour un processus stationnaire. Thesis at Univ. Sciences et Techniques, Lille 1, France.

Carbon, M. (1983) InAgalitA de Bernstein pour les processus fortement mAlangeants non nAcessairement stationnaires. C. R. Acad. Scienc., Paris, I, 297, pp. 303-306.

Castellana,J.V. and Leadbetter M.R. (1986) On smoothed probability density estimation for stationary processes. Stochastic Processes and their Applications, 21, pp. 179-193.

Cencov N.N. (1988) Why L_1 and what are at the horizon? (in Russian) Editorial comments to the Russian translation of Devroye and Györfi (1985a) Mir, Moscow.Y.S.

Chow, Y.S. (1967) On a strong law of large numbers for martingales. Annals of Math. Stat., 38, pp. 610.

Chanda, K.C. (1974) Stong mixing properties of linear stochastic process. Journ. of Appl. Proba., 11, pp. 401-408.

Collomb, G. (1976) Estimation nonparamAtrique de la régression par la methode du noyau. Thesis at University P. Sabatier, Toulouse, France.

Collomb, G. (1981) Estimation non paramétrique de la régression: revue bibliographique. Inter. Stat. Review, 49, pp. 75-93.

Collomb, G. (1983) From nonparametric Regression to nonparametric prediction: Survey on the mean square error and original results on the predictogram. Lectures Notes in Statistics, 16, pp. 182-204.

Collomb, G. (1984) Propriétés de convergence presque complète du predicteur à noyau. Z. Wahrscheinlichkeitstheorie und verw. Gebiete, 66, pp. 441-460.

Collomb, G. (1985a) Nonparametric regression: an up to date bibliography. Statistics, 2, pp. 309-324.

Collomb, G. (1985b) Nonparametric time series analysis and prediction: uniform almost sure convergence. Statistics, 2, pp. 297-307.

Collomb, G. and Doukhan, P. (1983). Estimation nonparametrique de la fonction d'autoregression d'un processus stationnaire et φ-melangeant: risques quadratiques par la mAthode du noyau. C.R. Acad. Sci. Paris, 296, I, pp. 859-862.

Collomb, G. and Härdle, W. (1986) Strong uniform convergence rates in robust nonparametric time series analysis. Stochastic Processes and their Applications, 23, pp. 77-89.

Collomb, G. Härdle, W. and Hassani, S. (1987) A note on prediction via estimation of the conditional mode function. J. Statist. Planning and Inference, 15, pp. 227-236.

Collomb, G., Hassani, S., Sarda, P. and Vieu, P. (1985) Est- -imation non-paramétrique de la fonction de hasard pour des observations dépendantes. Statistique et Analyse des données, 10, pp. 42-49.

Davydov, Y.A. (1968) Convergence of distributions generated by stationary stochastic process. Theory of Probability and its Applications, 13, pp. 691-696.

Davydov, Y.A. (1973) Mixing conditions for Markov chains. Theory of Probability and its Applications, 18, pp. 312-328.

Dehling, H. and Philipp, W. (1982) Almost sure invariance principles for weakly dependent vector valued random variables. Annals of Proba., 10, pp. 689-701.

Delecroix, M. (1980) Sur l'estimation des densités d'un processus stationnaire à temps continu. Publ. de l'Institut de l'Université de Paris, 25, pp. 17-40,

Delecroix, M. (1987) Sur l'estimation et la prévision non-paramétrique des processus ergodiques. Thesis at University of Lille Flandres Artois, Lille, France.

Devroye, L. (1981) On the almost everywhere convergence of nonparametric regression function estimates. Annals of Statistics, 9, pp. 1310-1319.

Devroye, L. (1982) Necessary and sufficient conditions for the pointwise convergence of nearest neighbor regression function estimates. Z. Wahrscheinlichkeitstheorie und verwandte Gebiete, 61 , pp. 467-481.

Devroye, L. (1982) Any discrimination rule can have an arbitrarily bad probability of error for finite sample size. IEEE Trans. on Pattern Analysis and Machine Intelligence, PAMI-4, pp. 154-157.

Devroye, L. (1983a) On arbitrarily slow rates of global convergence in density estimation. Z. Wahrscheinlichkeitstheorie und verwandte Gebiete,

61, pp. 457-483.

Devroye, L. (1983b) The equivalence of weak, strong and complete convergence in L_1 for kernel density estimates. Annals of Statistics, 11, pp. 896-904.

Devroye, L. (1987) A Course in density estimation. Birkhäuser.

Devroye, L. (1988) The kernel estimate is relatively stable Probability Theory and Related Fields (to appear)

Devroye, L. and Györfi, L. (1985a) Nonparametric density estimation: the L_1 view. Wiley.

Devroye, L. and Györfi, L. (1985b) Distribution-free exponential bound for the L_1 error of partitioning estimates of a regression function. In: Probability and Statistical Decision Theory, Proceedings of the Fourth Pannonian Symposium on Mathematical Statistics, F. Konecny, J. Mogyorodi, W Wertz Eds., Reidel, Dordrecht, pp. 67-76.

Devroye, L. and Krzyzak, A. (1987) An equivalence theorem for L_1 convergence of the kernel regression estimate. Report, McGill University, Montreal.

Devroye, L. and Wagner, T.J. (1980a) Distribution-free consistency results in nonparametric discrimination and regression function estimation. Annals of Statistics, 8, pp. 231-239.

Devroye, L. and Wagner, T.J. (1980b) On the L_1 convergence of kernel estimators of regression functions with applications in discrimination. Z. Wahrscheinlichkeitstheorie, 51, pp. 15-25.

Devroye, L. and Wise, G.L. (1980) Consistency of a recursive nearest neighbor regression function estimate. J. Multivariate Analysis, 10, pp. 539-550.

Doob, J. (1953) Stochastic processes. Wiley, New-York.

Doukhan, P. and Ghindès, M. (1980a) Estimation dans le processus $X_{n+1} = f(X_n)+\epsilon_n$. C. R. Acad. Scienc., Paris, 297, A, pp. 61-64.

Doukhan, P. and Ghindès, M. (1980b) Etude du processus $X_{n+1}=f(X_n).\epsilon_n$. C. R. Acad. Scienc., Paris, 290, A, pp. 921-924.

Doukhan, P. and Ghindès, M. (1980c) Etude du processus $X_n=f(X_{n-1})+\epsilon_n$. Thesis at Univ. Paris Sud, Orsay, France.

Doukhan, P. and Ghindès, M. (1983) Estimation de la transition de probabilitA d'une chaine de Markov Doeblin-rAcurrente. Stoc. Proc. and Appl., 15, pp. 271-293.

Doukhan, P. and LAon, M. (1988) Quelques notions de mAlange et des exemples de processus mAlangeants. Preprint, University Paris Sud, Orsay, France.

Farell, R.H. (1982) On the best obtainable asymptotic rates of convergence in estimation of a density function at a point. Ann. Statist., 43, 1, pp. 170-180.

Földes, A. (1974) Density estimation for dependent samples. Studia Scientiarium Mathematicarum Hungarica, 9, pp. 443-452.

Földes, A. and Révész, P. (1974) A general method for density estimation. Studia Scientiarium Mathematicarum Hungarica, 9, pp. 81-92.

Fritz, J. (1974) Learning from ergodic sample. In Limit Theorems of Probability Theories, P. Révész Ed., North Holland, pp. 79-91.

Fritz, J. (1975) Distribution-free exponential error bound for nearest neighbor pattern classification. IEEE Transactions on Information Theory, IT-21, pp. 552-557.

Gasser, T. and Müller, H.G. (1979) Kernel estimation of regression functions. in Smoothing techniques for curve estimation, ed. Gasser & Rosenblatt, Springer Verlag, Heidelberg, pp. 23-68.

Gasser, T. and Müller, H.G. (1984) Nonparametric estimation of regression functions and their derivatives by the kernel method. Scandinavian Journ. of Stat., 11, pp. 171-185.

Gorodetskii, V.V. (1977) On the strong mixing condition for linear sequences. Theory Proba. and Appl., 22, pp. 411-413.

Greblicki, W., Krzyzak, A. and Pawlak, M. (1984) Necessary and sufficient consistency conditions for a recursive kernel regression estimate. Annals of Statistics, 12, pp. 1570-1575.

Greblicki, W. and Pawlak, M.(1987) Necessary and sufficient consistency conditions for a recursive kernel regression estimate. J. Multivariate Analysis, 23, pp. 67-76.

Györfi, L.(1974) Estimation of probability density and optimal decision function in RKHS. In Progress in Statistics, J. Gani, K Sarkadi, I. Vincze Eds., North-Holland, pp. 281-301.

Györfi, L.(1976) An upper bound on error probabilities for multi-hypotheses testing and its application in adaptive pattern recognition. Problems of Control and Information Theory, 5, pp. 449-457.

Györfi, L.(1978) On the rate of convergence of nearest neighbor rules. IEEE Trans. on Information Theory, IT-24, pp. 509-512.

Györfi, L. (1981a) The rate of convergence of k-NN regression estimates and classification rules. IEEE Trans. on Information Theory, IT-27, pp. 362-364.

Györfi, L.(1981b) Strong consistent density estimate from ergodic sample. J. Multivariate Analysis, 11, pp. 81-84.

Györfi, L.(1981c) Recent results on nonparametric regression estimate and multiple classification. Problems of Control and Information Theory, 10, pp. 43-52.

Györfi, L.(1987) Density estimation from dependent sample. In Statistical Data Analysis Based on the L_1-norm and Related Methods Y. Dodge Ed. North-Holland, pp. 393-402.

Györfi, L. and Masry, E. (1988) The L_1 and L_2 strong consistency of recursive kernel density estimation from dependent samples. Report, TU Budapest.

Härdle, W. and Bowman, A. (1987) Bootstrapping in nonparametric regression: local adaptive smoothing and confidence bounds. J.A.S.A., 83, pp. 127-141.

Härdle, W., Hall, P. and Marron, J.S. (1988) How far are asymptotically chosen regression smooothers from their optimum? (with discussion) J.A.S.A., 83, pp. 86-101.

Härdle, W. and Luckhaus, S. (1984) Uniform consistency of a class of regression estimators. Annals of Statistics, 12, pp. 612-623.

Härdle, W. and Marron, J.S. (1985) Optimal bandwidth selection in nonparametric regression estimation. Annals of Statistics, 13, 4, pp. 1465-1481.

Härdle, W., Vieu, P. and Hart, J. (1989) Asymptotic optimal data-driven bandwidths for regression under dependence. Preprint.

Härdle, W. (1989) Applied Nonparametric Regression. Econometric Society Monograph Series, Cambridge University Press, in print

Hall, P. (1983) Large sample optimality of least squared cross- validation in density estimation. Ann. Statist., 11, pp. 1156-1174.

Hall, P. (1984) Asymptotic properties of integrated squared errors and cross-validation for kernel estimation of a regresssion function. Z. f. W. u v. G., 67, pp. 175-196

Hall, P. (1987a) On the use of compactly supported density in problems of discrimination. J. of Multiv. Anal., 23, 1, 131-158.

Hall, P. (1987b) On Kullback-Leibler loss and density estimation. Ann. Statist., 15, 4, pp. 1491-1519.

Hall, P. and Marron, J.S. (1987) On the amount of noise inherent in bandwidth selection for a kernel density estimate. Ann. Statist., 15, pp. 163-181.

Hall, P. and Marron (1988) Lower bounds for bandwidth selection in density estimation. North Carolina Institute of Statistics, Mimeo Series #1761.

Hall, P. and Schucany, W. (1988) A local cross-validation algorithm. Preprint.

Hart, J. (1988) Kernel regression estimation with time series errors.

Preprint.

Hart, J. and Vieu, P. (1988) Data-driven bandwidth choice for density estimation based on dependent data. Preprint

Hassani, S., Sarda, P. and Vieu, P. (1986) Approche non paramétrique en théorie de la fiabilité. Revue de Statistiques Appliquées, XXXV, 4.

Huber, P.J. (1983) Data Analysis: in search of an identity. Talk at the Neyman-Kiefer Symposiun in Berkeley, June 20, 1985.

Ibragimov, I.A.(1962) Some limit theorems for stationary processes. Theory of Probability and its Applications, 7, pp. 349-382.

Ibragimov, I.A. and Linnik, Y.V. (1971) Independent and stationary sequences of random variables. Gröningen: Wolters-Noordhoff.

Kolmogorov, A.N. and Rozanov, Y.A. (1960) On strong mixing conditions for stationary Gaussian processes. Theory of Probability and its Applications, 5, pp. 204-207.

Krzyzak, A. (1986) The rates of convergence of kernel regression estimates and classification rules. IEEE Trans. on Information Theory, IT-32, pp. 668-679.

Krzyzak, A. and Pawlak, M. (1983) Universal consistency results for Wolverton-Wagner regression function estimates with application in discrimination. Problems of Control and Information Theory, 12, pp. 33-42.

Krzyzak, A. and Pawlak, M. (1984a) Almost everywhere convergence of a recursive regression function estimate and classification. IEEE Trans. on Information Theory, IT-30, pp. 91-93.

Krzyzak, A. and Pawlak, M.(1984b) Distribution-free consistency results for Wolverton-Wagner regression estimate and classification. IEEE Trans. on Information Theory, IT-30, pp. 78-81.

Lejeune, M. (1985) Estimation nonparamétrique par noyau: régression polynomiale mobile. Revue de Statistique Appliquée, 33, 3, pp. 43-67.

Lejeune, M. and Sarda, P. (1988) Smooth estimators for density and distribution functions, a unifying approach. Preprint.

Loftsgaarden, D.O. and Quensenberry, C.P. (1965) A nonparametric estimate of a multivariate density function. Annals of Mathematical Statistics, 36, pp. 1049-1051.

Mack, Y.P. and Rosenblatt, M. (1979) Multivariate k-nearest neighbor density estimates. J. Multivariate Analysis, 9, pp. 1-15.

Mack, Y.P. and Silvermann, B.W. (1982) Weak and stong uniform consistency of kernel regression estimates. Z. Wahrscheinli- -chkeitstheorie und verw. Gebiete, 61, pp. 405-415.

Marron, J. S. (1987) A comparison of cross-validation techniques in density estimation. Ann. Statist., 15, pp. 152-162.

Marron, J.S. (1988) Automatic smoothing parameter selection: a survey. To appear in Econometric Reviews.

Marron, J.S. and Härdle, W. (1985) Random approximations of some measures of accuracy in nonparametric curve estimation. J. of Mult. analysis, 20, pp. 91-113.

Masry, E. (1983) Probability density estimation from sampled data. IEEE Trans. on Information Theory, IT-29, pp.696-709.

Masry, E. (1986) Recursive probability density estimation for weakly dependent stationary processes. IEEE Trans. on Information Theory, IT-32, pp. 254-264.

Masry, E. (1987) Almost sure convergence of recursive density estimators for stationary mixing processes. Statistics and Probability Letters, 5, pp.249-254.

Masry, E. and Györfi, L. (1987) Strong consistency and rates for recursive probability density estimators of stationary processes. J. Multivariate Analysis, 22, pp. 79-93.

McLeish, D.L. (1975) A maximal inequality and dependent strong law. Annals of Probability, 3, pp. 829-839.

Mielniczuk, J., Sarda, P. and Vieu, P. (1988) Local data-driven bandwidth choice for density estimation. To appear in J. of Statist. Planning and Inference.

Moore, D.S. and Yackel, J.W. (1977) Consistency properties, of nearest neighbour density functions. Annals of Statistics, 5, pp. 143-154.

Murthy, V.K. (1965) Estimation of jumps, reliability and hazard rate. Annals of Statistics, 36, pp. 1032-1040.

Nadaraya, E.A. (1964) On estimating regression. Theory of Probability and its Application, 9, pp. 141-142.

Parzen, E. (1962) On estimation of a probability density function and the mode. Annals of Mathematical Statistics, 33, pp. 1065-1076.

Peligrad, M. (1987) Properties of uniform consistency of the kernel estimator of density and of regression. Preprint.

Philipp, W. (1977) A functional law of the iterated logarithm for empirical distribution functions of weakly dependent random variables. Annals of Probability, 5, pp. 319-350.

Rejtö, L. and Révész, P. (1973) Density estimation and pattern classification. Problems of Control and Information Theory, 2, pp. 67-80.

Révész, P. (1972) On empirical density function. Periodica Mathematica Hungarica, 2, pp. 85-110.

Révész, P. (1973) Robbins-Monro procedures in a Hilbert space and its application in the theory of learning processes. Studia Sci. Math. Hung. 8, pp. 391-398.

Révész, P.(1979) On the nonparametric estimation of the regression function. Problems of Control and Information Theory, 8, pp. 297-302.

Rice, J. (1984) Bandwidth choice for nonparametric regression. Ann. Statist., 33, pp. 1215-1230.

Rice, J. and Rosenblatt, M. (1976) Estimation of the log survivor function and hazard function. Sankhya, A, 38, pp. 60-78.

Rosenblatt, M. (1956a) A central limit theorem and strong mixing condition. Proc. Nat. Acad. Sci., 42, pp. 43-47.

Rosenblatt, M. (1956b) Remarks on some nonparametric estimates of a density function. Annals of Mathematical Statistics, 27, pp. 832-835.

Rosenblatt, M. (1970) Density estimates and Markov sequences. In Nonparametric Techniques in Statistical Inrerence, M. Puri, Ed. London: Cambridge University, pp. 199-210.

Rosenblatt, M. (1971a) Curve estimates. Annals of Mathematical Statistics, 42, pp. 1815-1842.

Rosenblatt, M. (1971b) Markov Processes, Structure and Asymptotic Behavior. Springer Verlag.

Rosenblatt, M. (1979) Global measure of deviation of kernel and nearest neighbor density estimates. In Smoothing Techniques for Curve Estimation, Th. Gasser, M. Rosenblatt, Eds., Springer Lecture Notes on Math., pp. 181-190.

Rosenblatt, M. (1985) Stationary Sequences and Random Fields. Birkhauser.

Roussas, G.G. (1969) Non-parametric estimation of the transition distribution of a Markov processes. Annals of Inst. Statist. Math., 21, pp. 73-87.

Rudemo, M. (1982) Empirical choices of histogramms and kernel density estimates. Sacnd. Journ. of Statist., 9, pp. 65-78.

Sarda, P. and Vieu, P. (1985a) Régression nonparamétrique d'un processus Markovien. Cahiers du C.E.R.O., Brussell, 28, pp. 203-209

Sarda, P. and Vieu, P. (1985b) Nonparametric regression estimation, application to prediction. Proceedings of the 4th E.Y.S.M., Pliska, Studia Mathematica Bulgarica.

Sarda, P. and Vieu, P. (1988a) Vitesse de convergence d'estimateurs D noyau de la régression et de ses dérivées. To appear in C. R. Acad. Scienc., Paris.

Sarda, P. and Vieu, P. (1988b) Empirical distribution function for mixing random variables: Application in nonparametric hazard estimation. To appear in Statistics.

Sarda, P. and Vieu, P. (1989a) Smoothing parameter selection in hazard estimation. Preprint.

Sarda, P. and Vieu, P. (1989b) Estimation nonparamAtrique de la fonction de hasard. Preprint.

Scheffé, H. (1947) A useful convergence theorem for probability distributions. Annals of Mathematical Statistics, 18, pp. 434-458.

Scott, D.W. and Terell, G.R. (1987) Biased and unbiased cross-validation in density estimation. J.A.S.A., 82, pp. 1131-1146.

Shibata, R. (1981) An optimal selection of regression variables. Biometrika, 68, pp. 45-54.

Shields, P.C. (1973) The Theory of Bernouilli Shifts. The University of Chicago Press.

Silverman, B.W. (1986) Density estimation for statistics and data analysis. Chapmann and Hall, London.

Singpurwalla, N.D. and Wong, Y. (1983) Estimation of the failure rate. A survey of nonparametric methods. Part I: non Bayesian methods. Comm. in Stat. Th. Math., 12, 5, pp. 559-588.

Spiegelman, C. and Sacks, J. (1980) Consistent window estimation in nonparametric regression. Annals of Statis., 8, pp. 240-246.

Stone, C.J.(1977) Consistent nonparametric regression. Annals of Statist., 5, pp. 595-645.

Stone, C.J. (1982) Optimal global rates of convergence of nonparametric regression. Annals of Statist., 10, pp. 1040-1053.

Stone, C.J. (1983) Optimal global rate of convergence for nonparametric estimate of a density function and its derivatives. Recent Advances in Statistics. Paper presented in Honor of Herman Chernoff's sixtieth birthday.

Stone, C.J. (1984) An asymptotic optimal window selection rule for kernel density estimates. Ann. Statist., 12, pp. 1285-1297.

Truong, Y.K. and Stone, C.J. (1988a) Nonparametric time series prediction: kernel estimates based on local averages.Preprint.

Truong, Y.K. and Stone, C.J. (1988b) Nonparametric time series prediction: kernel estimates based on local medians. Preprint.

Vieu,P. (1988) Nonparametric regression: local optimal bandwidth choice. Preprint.

Watson, G.S. (1964) Smooth regression analysis. Sankhya A, 26, pp. 359-372.

Withers, C.S. (1981) Conditions for linear processes to be strong mixing.

Z. Wahrscheinlichkeitstheorie und verw. Gebiete, 57, pp. 477-480.

Wolverton, C.T. and Wagner, T.J. (1969) Asymptotically optimal discriminant functions for pattern classification. IEEE Trans. on Information Theory, IT-15, pp. 258-265.

Yakowitz, S. (1985) Markov flow models and the flood warning problem. Water Ressources research, 21, pp. 81-88.

Yakowitz, S. (1989) Nonparametric density and regression estimation for Markov sequences without mixing assumptions. Accepted for publication in Journ. of Multiv. Anal.

Yamoto, H. (1971) Sequential estimation of a continuous probability density function and mode. Bull. Math. Statist., 14, pp. 1-12.

Yokoyama, R. (1980) Moment bounds for stationary mixing sequences, Z. Wahrscheinlichkeitstheorie und verw. Gebiete, 52, pp. 45-57.

AUTHOR INDEX

SUBJECT INDEX